William Lees

Elements of Acoustics, Light, and Heat

William Lees

Elements of Acoustics, Light, and Heat

ISBN/EAN: 9783743392595

Manufactured in Europe, USA, Canada, Australia, Japa

Cover: Foto ©berggeist007 / pixelio.de

Manufactured and distributed by brebook publishing software
(www.brebook.com)

William Lees

Elements of Acoustics, Light, and Heat

IN COURSE OF PUBLICATION.

ADVANCED SCIENCE SERIES.

Adapted to the requirements of Students in Science and Art Classes, and Higher and Middle Class Schools.

Printed uniformly in 12mo, averaging 350 pp., fully Illustrated; cloth extra, price, $1.50 each.

Putnam's Elementary Science Series.

ELEMENTS

OF

ACOUSTICS, LIGHT, AND HEAT.

BY

WILLIAM LEES, M.A.,

LECTURER ON PHYSICS, WATT INSTITUTION AND SCHOOL OF ARTS, EDINBURGH;
LATE EXAMINER IN MATHEMATICS, UNIVERSITY OF EDINBURGH.

With Illustrations.

NEW YORK:

G. P. PUTNAM'S SONS,

FOURTH AVENUE AND TWENTY-THIRD STREET.

1873.

PREFACE.

THE following treatise has been prepared in strict accordance with the syllabus of the Government Department of Science and Art, as indicated by their scheme of instruction for the elementary stage examination, in the particular branches of Acoustics, Light, and Heat.

Having taught these subjects for several years to large classes, and that with encouraging success, as tested by the annual examinations of my students, I feel the less hesitation in complying with the request made to me by the Publishers, of giving an outline of the course which I have followed, in the hope that it may be found useful to others engaging in the discharge of similar duties.

While instruction in the branches referred to, or indeed in physical science generally, is essentially dependent for its thorough efficiency on extensive and minute experimental illustration, the possession on the part of the teacher of the requisite instruments and apparatus, is of course, a *sine quâ non*.

Special references, accordingly, are made to these, and such explanations given of them by means of diagrams

541789

and otherwise, as will enable the student to understand their construction and use, and thus aid him towards acquainting himself with the leading principles of these important departments of science.

At the end of each subject are added a few general questions, similar in some respects to those that have been given at the May examinations.

In an Appendix I have made a selection of questions from some of the former Government papers, and have given also their solutions. These, it is to be hoped, will be of use to the student in the way of showing him how to set down his knowledge of the subjects for the examiner.

Though many excellent books on Physics have been written of late years, both in our own country and on the continent, I must own myself more especially indebted to the works of Tyndall, Ganot, and Deschanel. For a more extensive and complete knowledge of the subjects in question, the student would do well to refer to these works.

W. L.

Linkvale Lodge, Viewforth,
Edinburgh, *December*, 1872.

CONTENTS.

---◆◇◆---

ACOUSTICS.

CHAPTER I.

LIGHT.

CHAPTER I.

CHAPTER II.

CHAPTER III.

CHAPTER IV.

CHAPTER V.

HEAT.

CHAPTER I.

CHAPTER II.

CHAPTER III.

CHAPTER IV.

CHAPTER V.

CHAPTER VI.

CHAPTER VII.

APPENDIX.

FORMER EXAMINATION QUESTIONS, WITH THEIR SOLUTIONS.

ACOUSTICS.

CHAPTER I.

1. Object of Acoustics.—The term "acoustics" is derived from a Greek verb signifying "to hear." It is applied to designate that branch of science which treats of the phenomena of sound.

2. Cause of Sound.—The *immediate* cause of sound is the vibration of the sounding body. If, for instance, we take a glass receiver, and holding it by the top, strike it with a wooden mallet, it emits a clear ringing sound; and we can be assured of the fact that it is in a state of vibration, by observing the tremulous motion of the mallet when allowed to rest lightly on the side of the receiver—or by suspending a series of cork balls from the top of the receiver, when a peculiar dancing motion of the balls takes place.

3. How the Air is Affected—Amplitude.—The question arises, in what way is the air affected by these vibrations on the part of the sonorous body? The particles of air in the immediate vicinity of the body are thrown into a forward, and thence by their elasticity into a backward motion, passing to a short distance, then returning, and so on successively; but the air contiguous to this directly affected portion of air takes up the impression, and a similar motion of the aërial particles takes place; in like manner the air contiguous to this second affected portion takes up the impression; and thus the original motion is transmitted from one portion of air to another, a sonorous wave. The wave gradually enlarges as it leaves the bell, whilst at the same time the motion of the air particles becomes less and less, just as in the case of the concentric rings which are observed when a stone is dropped into a pool of still water.

The sounding body, whatever that be, is continually sending out a succession of such waves. These sound-waves enter our ears, affect the auditory nerve, and produce the sensation which we call "sound."

5. Sound is not Transmitted through a Vacuum.—It is essential that there be some medium for the transmission of sound. The ordinary channel of conveyance is the air; but, as we shall afterwards see, there are other substances which convey sound. That sound cannot be transmitted through a vacuum is proved by the ordinary experiment of ringing a bell under the receiver of an air-pump. As the air becomes exhausted, the sound of the bell gradually diminishes in intensity till it becomes almost inaudible. The sound indeed would be quite inaudible were it possible to make a complete vacuum, and were we able to dispense with the supports of the bell.

This experiment proves also that the intensity of sound diminishes with the density of the air—a fact which is well known to those who ascend lofty mountains. Thus it is said that at the top of Mont Blanc the human voice is much weakened, and that the report of a pistol resembles the noise of a boy's pop-gun. It follows from such observations, that the loudest noises or explosions which take place on the earth's surface, could not be heard beyond the limits of the atmosphere.

6. Velocity of Sound—How Determined—Sound is not conveyed from one place to another instantaneously—time is required for its propagation. This is abundantly evident from our most familiar observation. We hear the blows of a hammer at a distance some appreciable time after it has struck the object; the report of a distant gun reaches our ears some time after we see the flash.

The velocity of sound through air, at a given temperature, has been determined in the following manner:—

A ———————————————————————————————————— B

Fig. 2.

Let the distance between two stations, A and B, be carefully measured. Let a party at A (fig. 2) fire a gun, whilst another party at B counts the *number* of seconds between seeing the flash and hearing the report. The rate at which the sound travels per second will therefore be found by dividing the distance A B by that number.*

From a series of very careful experiments made in different countries, it appears that the velocity of sound at the *freezing* point may be taken at 1090 *feet per second,* and that the *increase* of velocity for every single degree of the centigrade thermometer amounts to nearly *two* feet.

Solution of Questions.—There are certain questions connected with the velocity of sound and the temperature, in which the student would do well to exercise himself. The nature of these will be understood from the following examples :—

Ex. 1.—*Find the velocity of sound through air, when the temperature is* 25° C.

For every degree centigrade there is an increase of 2 feet in the velocity; therefore for 25°, *an increase of* 25 × 2 *or* 50 *feet; hence the velocity at* 25° C. = 1090 + 50 = 1140 *feet.*—Ans.

Ex. 2.—*Given the velocity of sound to be* 1120 *feet per second, what is the temperature of the air?*

Here, if we deduct 1090 *from the given velocity, we obtain the amount of increase above the velocity at* 0° C.; *hence the temperature* = 30 ÷ 2 = 15° C.—Ans.

---

* The method, it will be observed, is founded on the *instantaneous* passage of light—a doctrine which is not, strictly speaking, true. Light travels, however, so fast, that for ordinary distances the time required for its transmission is virtually inappreciable.

Ex. 3.—*An interval of* 3½ *seconds is observed to elapse between a flash of lightning and the peal of thunder, what is the distance of the electric cloud, the temperature of the air being* 30° C.?

By the method of Ex. 1, *we find the velocity of sound to be* 1150 *feet; hence the distance* = 1150 × 3½ = 4025 *feet.*—Ans.

7. Elasticity and Density—Influence of Temperature.—The two conditions that must be taken into account in determining the velocity of propagation of a sonorous wave through any medium are its *elasticity and density.* Speaking more strictly, it is *the relation the former bears to the latter,* which determines the velocity of propagation. It is proved mathematically that the velocity is directly proportional to the square root of the elasticity, and inversely proportional to the square root of the density.

Now, in regard to the atmosphere, an increase of temperature causes a decrease in the density of the air; and a decrease in the temperature causes an increase in the density, the elasticity remaining the same. In the former case, therefore, sound travels *faster,* and in the latter case *slower.*

If the temperature, however, remain constant—according to Boyle's or Marriotte's law *—the elasticity varies in the same proportion as the density; and since these conditions are directly opposed to each other, it follows that the velocity of sound through air is, on this supposition, in no way affected. Consequently, there is no effect on the velocity of sound, unless change of density be accompanied by a change of temperature.

* Boyle's or Marriotte's law is generally enunciated thus:—
"The temperature being the same, the volume of a mass of air is inversely as the pressure it supports." Thus, if we have a mass of air occupying 1 cubic foot, at the ordinary pressure of the atmosphere (*i.e.,* 30 inches of mercury)—under a pressure of 2 atmospheres it will occupy ½ a cubic foot, under 10 atmospheres $\frac{1}{10}$ cubic foot, and so on. It follows, of course, that the elasticity or pressure is in direct proportion to the density.

It is from such considerations that we see the reason why the velocity of sound is not affected by changes in the barometer, but only by changes in the thermometer.

8. Changes of Temperature in a Sonorous Wave.— When air is compressed heat is evolved, and when rarefied cold is produced; therefore, "in the condensed portion of a sonorous wave the air is above, whilst in the rarefied portion of the wave it is below its average temperature. This change of temperature, produced by the passage of the sound-wave itself, virtually augments the elasticity of the air, and makes the velocity of sound about one-sixth greater than it would be if there were no change of temperature." * The temperature of the general mass of air, however, through which the sound-waves pass is *not* affected by these changes. Hence, in a concert room the air is not heated by the passage of the numerous sonorous waves which are constantly proceeding from the orchestra.

CHAPTER II.

9. Intensity of Sound.—We have already seen that the intensity of sound depends upon the density of the air (Art. 5). There are other things which may be noticed in regard to it.

(1) The intensity varies inversely as the square of the distance. This is known as *the law of inverse squares,* and is the same as regards light and radiant heat (see Art. 32).

(2) The intensity depends upon the density of the air in which the sound is generated, and not on that in which it is heard. Thus, if two observers be stationed at A and B, one on the top of a mountain and the other on the plain below, at equal distances from a gun at G (fig. 3),

---

* Tyndall on *Sound,* p. 46.

the report of the gun will have the same loudness to each, though the air at A is more rarefied than at B.   If, how-

Fig. 3.

ever, there be two guns, giving at the *same* place reports of equal loudness, then when placed at A and B respectively, to an observer stationed at C the gun at B will give a louder report than the gun at A.

(3) The intensity depends upon the amplitude (Art. 3). The relation between the two is more strictly expressed thus : *The intensity is proportional to the square of the amplitude.*

10. Propagation of Sound through Other Media.— Sound is not only transmitted through gaseous bodies, but also through liquids and solids.   The velocity through gases is comparatively small; it is greater through liquids, and still greater through solids.

The common notion that the velocity through a medium is in proportion to its density is quite erroneous.   We must take into account the relation which the elasticity of the medium bears to its density (Art. 7), and not either element in itself.   Thus, for example, if we take the gases, hydrogen and carbonic acid, we find that, in comparison with the velocity through air, sound travels *faster* through hydrogen and *slower* through carbonic acid —though the former has a less density, and the latter a greater density than air.   The true reason of the difference is, therefore, that in hydrogen its elasticity as compared with its density is greater than in the case of air, and in carbonic acid less.

In liquids, this relation is higher than in air; and in solids, still higher. Hence the velocity of sound is greater in the former, and still greater in the latter.

The following table gives the velocities through certain substances :—

VELOCITY OF SOUND THROUGH DIFFERENT SUBSTANCES—(Ft. per Sec.).

GASES.	LIQUIDS.	SOLIDS.
		Ash,...15314
Air,........ 1090(0°C)	Water(fresh), 4708 (8°C)	Oak, ..12622
Oxygen,.. 1040(0°C)	Alcohol, ...4218 (20°C)	Gold, 5717 (20°C)
Hydrogen 4164(0°C)	Solution of ⎱ 5132 (18°C)	Silver, 8553 ,,
Carb. Acid, 858(0°C)	Com. Salt ⎰	Iron,..16822 ,,

It appears from this table that the velocity through water is more than four times, through oak eleven times, and through iron fifteen times the velocity through air.

The readiness with which solids transmit sound is illustrated by several familiar facts. The slightest scratch with a pin at one end of an iron bar is distinctly heard at the other. By placing one end of a stick on the lid of a boiling kettle, and the other close to the ear, the commotion produced in the water is rendered very audible. The approach of a body of cavalry at a distance can be heard, it is said, by applying the ear to the ground.

11. **Reflection of Sound.**—When a wave of sound meets an obstacle in its course it is reflected. The law which regulates this reflection is the same as that of light (see Art. 36).

The efficiency of tubes to convey sound depends greatly on reflection. Biot made a number of interesting experiments with the water pipes at Paris. He found that with a tube, upwards of 3000 feet in length, the lowest whisper could be heard. "I wished," says Biot, "to ascertain the lowest pitch at which the voice ceased to be heard; but I could not succeed. Words spoken as low as when one whispers into the ear of another were conveyed and appreciated, and I concluded that the only possibility of not being heard was not to speak at all."

He found also that sounds of different pitch were conveyed with precisely the same speed.

The reflection of sound may be well illustrated by the following experiment:—

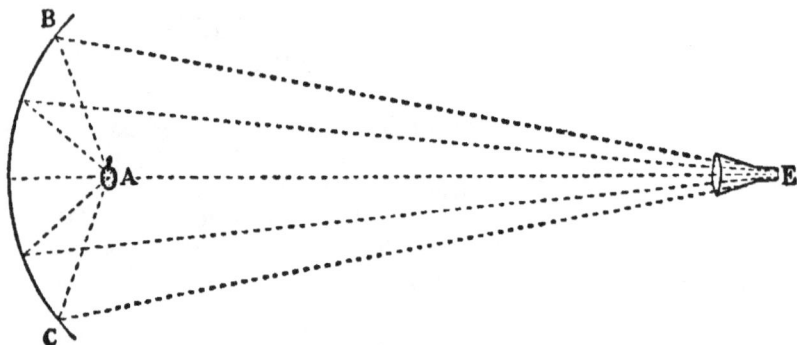

Fig. 4.

B C is a curved metallic reflector (fig. 4) placed behind a watch at A. If a person at E so adjusts himself as to have his ear in the focus of the sound-waves, the ticking of the watch will be distinctly heard. A funnel-shaped tube, as serving to entrap as many waves as possible, is of advantage in the experiment.

The ear-trumpet, the speaking-trumpet, whispering galleries, etc., all owe their action to the reflection of sound. Among the ancients a notorious instance of a sound-collecting surface was the " Ear of Dionysius" in the dungeons of Syracuse. The roof of the prison was so constructed as to transmit the words and even the whispers of the unhappy victims there confined, to the ear of the tyrant through a narrow passage cut through the solid rock.

12. Echoes.—An echo is also a result of the reflection of sound. But it becomes necessary to enquire into the precise circumstances under which it is produced.

For this purpose we must keep in view two things:
(1) the velocity of sound; and (2) the fact that the ear
cannot appreciate two sounds distinctly, unless an interval
of one-tenth of a second elapse between them. Now,
taking the standard temperature of 62° Fahrenheit, sound
travels at the rate of 1125 feet per second; in one-tenth of
a second, therefore, it will travel over 112½ feet. Hence
the *least* distance at which a person must place himself
from the reflecting surface, so as to make an echo possible,
must be about 56 feet. Within this distance the reflected
sound blends with the original, causing a certain amount
of enforcement, but no echo. On the other hand, as this
distance is exceeded, the echo becomes more and more
distinct.

Echoes are met with in all parts of the world, differing
much in their distinctness and character. One of the
best echoes to be found in Britain is in Woodstock Park.
It is said to repeat seventeen syllables by day and twenty
by night. At a villa near Milan there is one which
is said to repeat a shout thirty times.

The repetition which is met with in such cases results
from the repeated reflection of the original sound by
different obstacles, each repetition becoming fainter and
fainter, till it dies away in the distance.

The rolling of thunder is undoubtedly due in part to
the echoes caused by the presence of prominent objects,
such as houses and mountains, and to reflection amongst
the clouds. It is also believed to be owing to the different

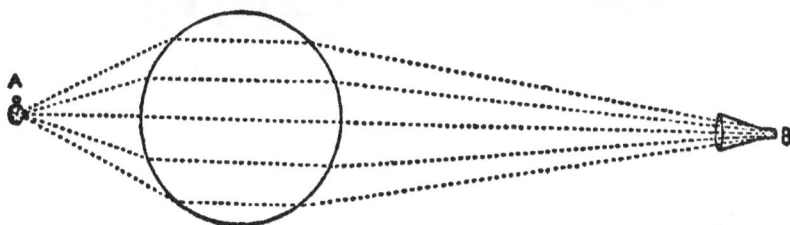

Fig. 5.

aërial agitations in the track of the lightning reaching the
ear at different times.

13. Refraction of Sound.—Sound, like light, can also be refracted, that is, bent out of its course by the interposition of some medium. This is effected in the case of sound by filling a thin india-rubber balloon with carbonic acid gas, and placing a watch, for example, at A, as in fig. 5. The sound-waves are refracted by the balloon, and concentrated in a focus at B, where the ticking of the watch may be distinctly heard.

14. Structure of the Ear—Range of Appreciation for Musical Sounds.—The human ear may be described as consisting of three parts—the *outer* ear, the *middle* ear, and the *labyrinth*. The accompanying diagram exhibits the different parts:—

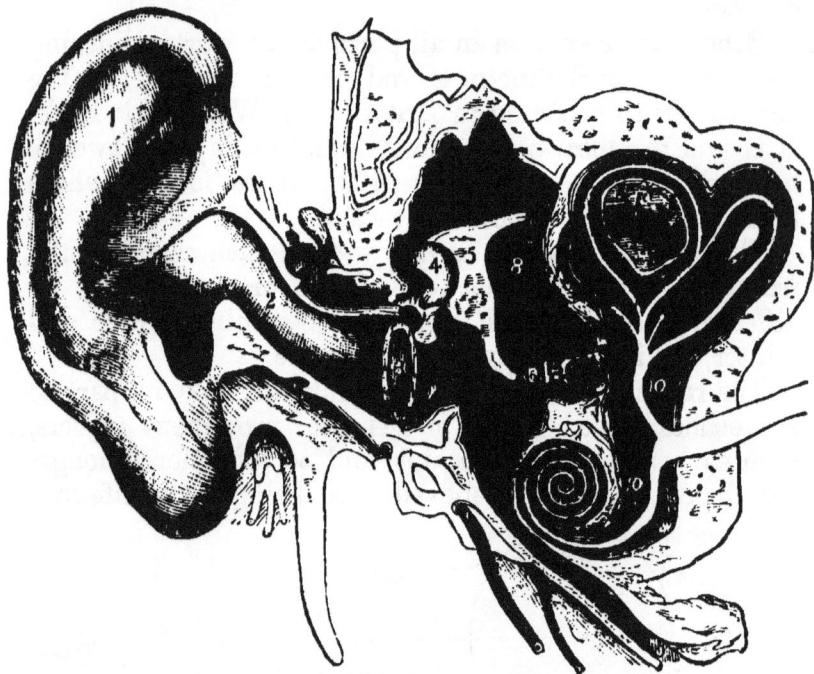

Fig. 6.

1—The *concha*. 2—The *meatus*. 3—The *tympanum* or *drum;* this closes the *outer* ear. 4, 5, 6, 7—A series of bones which transmit the impressions made on the tympanum, called respectively the *malleus, incus, os orbi-*

cularis, and *stapes.* 8, 8—The *tympanic cavity* or *middle*
ear. 9—The *Eustachian tube,* leading into the back of
the mouth, by which the air in the tympanic cavity is
kept of the same density as that of the external atmosphere, giving the tympanum therefore perfect freedom
of motion. 10, 10, 10—The *labyrinth,* throughout which
the auditory nerve is distributed.

The sonorous waves, entering the outer ear, throw the
tympanic membrane into a state of vibration. These
vibrations are transmitted across the middle ear through
the series of bones towards the labyrinth; there they
affect the auditory nerve, and produce the sensation of
hearing.

An ordinary musical ear can appreciate sounds arising
from 16 vibrations* per second up to 38,000, that is,
a range of about 11 octaves.† How is this accommo-
dation of the organ effected? Looking to the anatomy
of the tympanum, it appears that this adaptation to
different rates of vibration is effected by means of
slender muscles, which tighten or slacken the membrane
according as the vibrations which fall upon it are quick
or slow, thereby tuning, it, as it were, to the proper
discharge of its wonderful office.

CHAPTER III.

15. **Physical Difference between a Musical Sound
and Noise.**—The sensation we experience at once indicates the difference between a musical sound and noise.
But what is the real physical cause of the difference? It
is this: a musical sound is produced by *periodic* impulses

---

* By a vibration in this country is meant an excursion of the
vibrating body to *and* fro, not a movement backward or forward,
but both together.

† The practical range of musical sounds is from 40 to 4000
vibrations per second.

given to the air, that is, by impulses which succeed each other after perfectly regular intervals of time. A noise, on the other hand, is produced by impulses which do not succeed each other regularly. The vibrations of a tuning-fork, and the confused mingled noise of the street, are familiar illustrations.

It is sometimes difficult, however, to draw the exact line of demarcation, as, for example, in the case of a boy running a stick along a railing, or in the clink of the wheels of a railway carriage when it is moving rapidly.

16. Pitch, Intensity, and Quality of Musical Sounds. —Some sounds are said to be "grave," others "acute." This difference in the character of musical sounds depends upon what is called *pitch*, that is, *upon the number of vibrations performed in one second*. Thus, if we have two tuning-forks giving respectively 280 and 420 vibrations per second, the note from the former is of a lower pitch than the note from the latter.

Intensity arises from difference in the amplitude of the vibrations. The same sound or note of a certain pitch may have different degrees of loudness or intensity, according to the range of vibration given to the sounding body.

Quality (perhaps better expressed by the French word *timbre*), is the distinction which may be drawn between two notes of the same pitch and intensity, when sounded on different instruments, as, for example, on the violin and flute. It is believed to be due to certain subsidiary notes or "harmonics," as they are termed, accompanying, or being blended with, the original note.

17. Method of Determining the Number of Vibrations.—One of the simplest methods of determining the number of vibrations of a musical sound, is by means of Savart's apparatus. The machine consists of two wheels, A and B, fixed in a wooden frame (fig. 7), the smaller having a certain number of teeth in the rim. They are so adjusted, that B is made to revolve with great rapidity, its teeth hitting upon a card E fixed near it. The number of revolutions is indicated by a *counter* attached

to the axis at H. The method of procedure will be
understood by an example. The number of vibrations

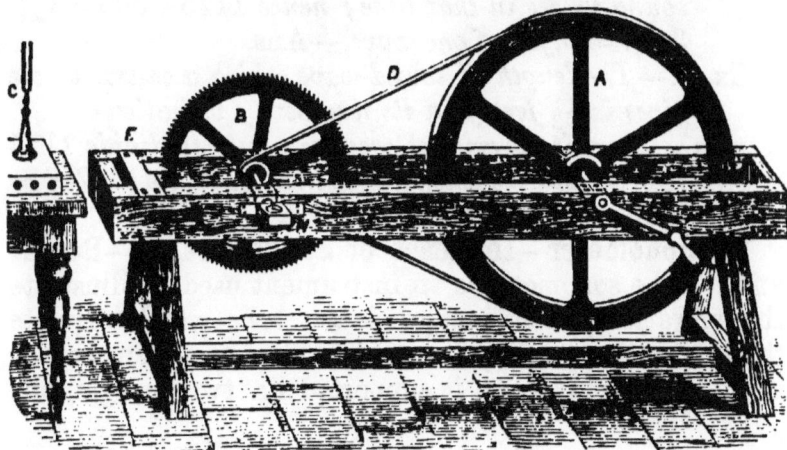

Fig. 7.

of a tuning-fork, C, mounted upon a sounding-box, is
required. We should first sound the fork, and gradually increase the revolution of the wheel, B, until the
note emitted by the card corresponds to that of the fork.
We should then keep them in unison for a certain number
of seconds, say ten. Now, if we suppose that there are
100 teeth in the wheel, B, and that during the ten seconds
the counter indicates fifty revolutions, we shall then have
5000 as the number of shocks or vibrations given to the
card in that time. Hence 5000 divided by 10, or 500,
will be the number of vibrations which the tuning-fork
performs per second.

Solution of Questions. — Given the number of
vibrations of a musical note, we can easily find the
length of sound-wave it produces. Conversely, given the
length of sound-wave, we can find the number of vibrations.

Ex. 1.—*A musical note gives* 300 *vibrations per second,
find the length of the sound-wave it produces.*

Taking the velocity of sound at 62° *Fahrenheit,
we have the first sound-wave sent off, travelling*

over 1125 *in one second; but the note in the example gives* 300 *vibrations, or sends off* 300 *sound-waves in that time;* hence 1125 ÷ 300 = 3 *ft.* 9 *in.* = *length of one wave.*—Ans.

Ex. 2.—*The length of sound-wave which a musical note gives is* 4½ *feet, find its number of vibrations.*

Here it is evident we have only to divide 1125 *feet by the wave-length;* hence 1125 ÷ 4½ = 250 = *number of vibrations.*—Ans.

19. Sonometer—Influence of Sound-Boards—Resonance.—The sonometer is an instrument used to illustrate the vibrations of strings, and the laws which regulate these vibrations. A convenient form of it is represented in fig. 8. Each string is supported on two bridges, one

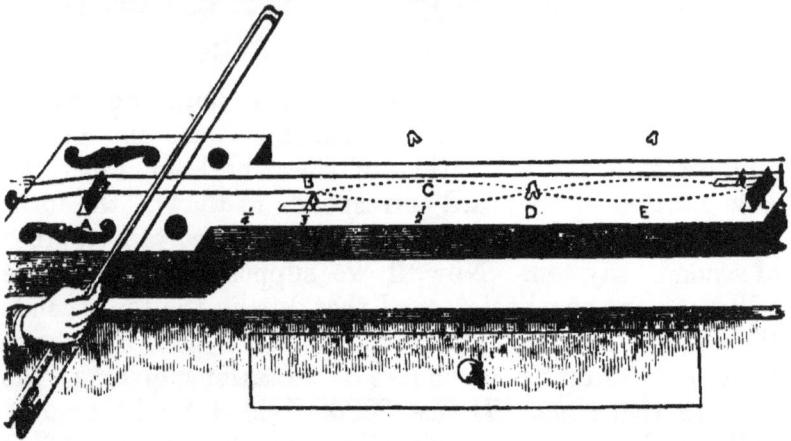

Fig. 8.

of which is movable, and by means of which, therefore, any part of the strings can be sounded. At one end there is a hollow box or sound-board A, which resounds under the influence of the vibrating strings, and thus very much enforces the sound. We have a similar effect in all our stringed instruments—violins, harps, pianos, etc.; they owe their richness and fulness of tone to the reverberation of the materials which support the strings. This effect is known as *resonance.*

We may have resonance also from a column of air.
Thus, if a tuning-fork be
held over the mouth of a
glass jar (fig. 9), the sound
is intensified. The height
of the jar, which gives the
maximum resonance, is
found by experiment to
be one - fourth of the
length of the sound-wave
which the fork pro-
duces.

**20. Nodes and Ventral
Segments.**—If a jerk or
pulse be sent along a string
fixed at one end, it is re-

Fig. 9.

flected at the fixed end, and returns to the hand. The time
it requires to travel to the fixed end and back, is the same
as that required by the whole string to execute a complete
vibration. If a series of jerks be sent in succession along
the string, the direct and reflected pulses meet, and by
their coalescence divide the string into a series of vibrating parts, called *ventral segments,* which are separated
from each other by points of apparent rest, called *nodes.*
This may be well shown by taking a few yards of common
window cord, as flexible as possible, fixing it at one end,
and after a succession of jerks has been given to it at the
other end, tightening the string slightly. It may divide
itself, as in fig. 10. A B, B C, C D are the ventral segments, B and C the nodes.

Fig. 10.

The number of ventral segments depends upon the
rapidity with which the jerks have been imparted.

This division of a string may also be illustrated by the

sonometer. Thus, if we place the bridge at a point, such
that A B is one-third of the length of the string (fig. 8),
and draw the bow across that part, the remainder of the
string will divide itself into two parts; that is, there will
be two ventral segments separated by a node. A simple
method of proving this is to take three small pieces of
cardboard, and place them at the points C, D, E. When
A B is sounded, the middle one will remain comparatively
unaffected, whilst the other two will be so much disturbed
by the vibration as to be thrown off.

21. **Laws of the Vibration of Strings.**—The transverse
vibration of strings is dependent upon four things; viz.,
length, thickness, tension, and density. Accordingly,
there are four laws which regulate the vibrations. They
are as follow :—

(1) *The number of vibrations of a string is inversely as
its length.*

(2) *The number of vibrations is inversely as its diameter.*

(3) *The number of vibrations is directly proportional
to the square root of the stretching force.*

(4) *The number of vibrations is inversely proportional
to the square root of its density.*

Hence half of a string gives double the number of vibrations—the note it produces is the octave to the original.
Hence, also, if we have two similar strings, the one twice
as long as the other but only half the diameter, these
strings will give the same note.

22. **Nodal Lines in a Vibrating Plate.**—We have
not only nodal points in a vibrating string, but we may
have *nodal lines* in a vibrating plate.

Thus, if we take a metallic plate (fig. 11), and sprinkle
sand uniformly over it, by drawing a fiddle-bow across
the middle point of one of its sides, the sand collects
along the diagonal lines, completely leaving the other
parts of the plate, which we infer are thus thrown into a
state of rapid vibration. If the bow be drawn across a
point near one of the corners, the sand is arranged

along the lines joining the middle points of the opposite sides.

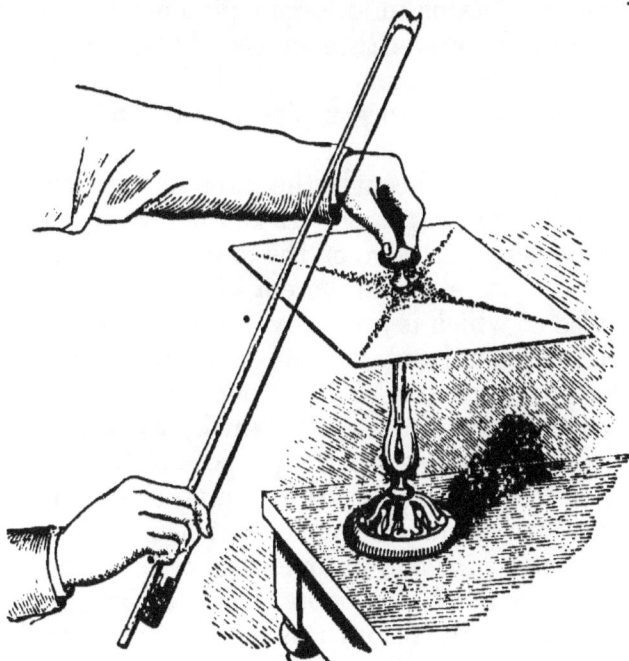

Fig. 11.

23. Stopped and Open Pipes.—We have now to look at the vibration of columns of air in pipes. It is the column of air itself, in such cases, which is the cause of the sound, and not the material of which the pipe is constructed.

In both stopped and open pipes the number of vibrations is inversely proportional to the length of the pipe. This may be illustrated by taking three glass tubes (fig. 12), A, B, C, say 16 in., 8 in., and 4 in. long, respectively, and blowing across their open ends.

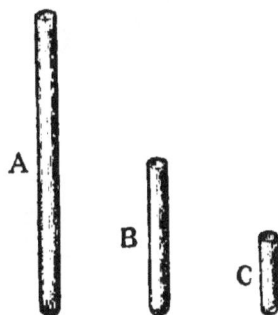

Fig. 12.

Whilst A gives a certain note, B gives the octave, and

C the octave to B. In the case of open tubes, it is difficult to elicit the notes in this manner, and an arrangement such as obtains in an organ pipe must be adopted.

**24. Organ Pipes—State of the Air in Stopped and
Open Pipes.**—The manner in which the column of air is
made to vibrate in an organ pipe will be understood from
fig. 13.

The air, urged through the tube A, is led
through the narrow passage A B, and made to
play upon the thin edge of the pipe at the *em-
bouchure* C. It produces there a kind of flutter,
some pulse of which is raised by the resonance of
the aërial column inside to a musical sound, and
thus the pipe "speaks." In an open pipe there
is a flexible metallic tongue, which, by being
moved up or down, serves the purpose of tuning
it. In a stopped pipe there is a plug or piston
at the top, which may be moved out or in.

If a stopped pipe and an open one of the same
length be sounded, the note emitted by the
former is an octave below the latter; hence, to
have the same note with a stopped pipe as with
an open one, the former must be half the
length of the latter.

When a stopped pipe gives its fundamental
note,* *the column of air inside is undivided by
a node—the stopped end is a node and the open
end the middle of a ventral segment.* In an open
pipe, again, *the column is divided by a node at
its centre—each end being the middle of a ventral
segment.*

The length of the sonorous wave of the Fig. 13.
note produced in a stopped pipe is *four* times the length
of the pipe, and in an open one *twice* the length.

* By the fundamental note of a pipe is meant the lowest note
which can be drawn from it. Harmonics may be drawn from a
pipe by increasing the intensity of the current of air urged
through it.

If hydrogen and carbonic acid gas be urged in succession through an organ pipe, the note produced in the former case is of *higher* pitch, and in the latter case of *lower* pitch, than when air is urged through the pipe. This results from the comparative velocities of sound in these gases.

25. Interference of Sound Waves—Beats in Music. —If a tuning-fork, which is held over the mouth of a resonant jar, be gradually moved towards the side its sound becomes enfeebled, and almost disappears, when in the position represented in fig. 14. This is due to the sound-waves proceeding from the prongs of the fork neutralizing each other — an effect known as *interference.* A similar effect takes place if the fork be held close to the ear, and slowly turned on its axis.

Fig. 14.

When two sonorous bodies, whose periods of vibration slightly differ, emit sound together, a series of alternate reinforcements and diminutions of the sound are observed, called in music *beats.* They are caused by the sonorous waves of the two bodies at one time conspiring, and at another time being opposed, in their action on each other—a result, therefore, of interference.

These beats may be well heard by sounding a low note and its sharp on a piano.

It may be proved that the number of beats per second is just equal to the difference between the rates of vibration of the two notes. Concord or harmony in music results from the frequency in the coincidence or coalescence of the vibrations; discord, on the other hand, from the unfrequency with which this occurs.

26. The Voice—Stuttering.—The top of the windpipe or *trachea* is closed by two membranes placed edge to edge, leaving only a small slit or opening between them, called the *glottis*. These membranes are acted upon by elastic bands, called the *vocal chords*, which are relaxed or tightened at will. There is a flap or lid, termed the *epiglottis* (fig. 15), which accurately covers the glottis.

The food, in its passage into the gullet, presses upon this lid and keeps it close upon the aperture until the food has passed; when, from its elasticity, it rises and allows respiration to go on. When we are breathing, but not speaking, the membranes of the glottis are in a state of relaxation, and the air in its exit from the lungs has too little force to cause them to vibrate; in these circumstances there is no sound, no voice. When we wish to speak, no sooner does the volition exist than the vocal chords brace up the membranes to the necessary tension; the lungs then doing their duty send a blast of air through the glottis, throw the membranes into vibration, and thus sound or voice is produced.

Fig. 15.

"The sweetness and smoothness of the voice depend on the perfect closure of the slit of the glottis, at regular intervals, during the vibration. . . . Through the agency of the mouth we can mix together the fundamental tone and the overtones (harmonics) of the voice in different proportions, and the different vowel sounds are due to different admixtures of this kind." *

Stuttering in speech is believed to arise from some inability on the part of the stutterer to open his glottis. It has been recommended to such persons when speaking

* Tyndall on *Sound*, pp. 196-199.

to drawl out their conversation in a continuous sound, so as to prevent their glottis from closing. Hence we find that stutterers can often sing well, and without the least interruption; for the tune being continued the glottis does not close, and there is consequently no hesitation. Many stutterers can also read poetry well, or any declamatory composition in which the sound is almost as much sustained as in singing.

QUESTIONS.

1. What is the velocity of sound in air at the freezing point? Calculate the velocity when the temperature is 10°C.
Ans. (1) 1090; (2) 1110.
2. How is the velocity of sound affected (1) by a change of density? and (2) by a change of elasticity? Why is the velocity through a solid greater than through a liquid, and still greater than through a gas?
3. Describe a method of determining the number of vibrations of a sonorous body.
4. Describe an experiment to prove that sound cannot pass through a vacuum.
5. Distinguish between a musical sound and a noise. On what do the pitch, intensity, and quality of musical sounds depend?
6. A sonorous body gives wave-lengths of $3\frac{1}{2}$ feet, how many vibrations does it execute in one second (temp. $=15°C$.)?
Ans. 320.
7. A and B are two similar strings stretched with the same force; A has twice the length of B, but only half the diameter. Compare their rates of vibration.   *Ans.* Equal.
8. What is the length of the sonorous wave produced (1) by a stopped pipe 2 feet long, and (2) by an open pipe of the same length, each sounding its fundamental note?
Ans. (1) 8 feet, (2) 4 feet.
9. How does an organ pipe "speak?" What is the state of the air inside (1) an open, and (2) a stopped pipe, when each gives its fundamental note.
10. What is the cause of *beats* in music? How many beats per second would occur when two tuning forks are sounded together, which give 300 and 310 vibrations per second respectively?

LIGHT.

CHAPTER I.

27. Theories in Regard to Light.—There are two main theories which have been proposed in reference to the nature of light. One of these is termed the *emission* or *corpuscular* theory, the other, the *undulatory* theory.

According to the former, light consists in the actual emanation of luminous particles from the luminous body; that these particles are of inconceivable minuteness, and by their actually striking upon our eyes, that they excite the sensation of vision, in the same way as the fine particles of any odorous substance, by entering our nostrils, excite the sense of smell.

In the latter hypothesis, it is assumed that all bodies, as well as all space, are pervaded with an extremely thin and highly elastic fluid.called *ether ;* that light is caused by a vibratory motion imparted to the molecules of this substance, in consequence of which a series of undulations or waves are produced; and that it is by the successive shocks of these minute waves upon our eyes that vision is excited, somewhat in the same manner as the waves of sound, proceeding from a sonorous body, by entering our ears, excite the sensation of sound.

The undulatory theory, from its perfect competency to explain all the varied phenomena connected with light, is the one more generally adopted by modern physicists.

28. Light Proceeds in Straight Lines—Definitions.— That light is propagated in straight lines is manifest from various facts. We cannot see round a corner. If we hold an opaque object in front of a candle, we fail to see the candle. If a small hole be made in the shutter of a darkened room, the track of a beam of light, as marked out by the floating particles of dust, is observed to be perfectly straight.

Towards sunset, when the sun is concealed by a cloud (fig. 17), straight beams of light are observed to radiate from him in all directions, and to shed a glowing effect upon a surrounding landscape.

Fig. 17.

A *ray* of light is an indefinitely small portion of light —a mere line in thickness. An assemblage of such rays is termed a *pencil* or *beam* of light; when the rays proceed in parallel directions, the pencil is said to be *parallel;* when they proceed in all directions, it is *divergent,* and when they converge towards a point, it is *convergent.* Parallel and convergent beams are met with in optical instruments; divergent beams are the most common, and are such as proceed from any luminous body.

29. Inversion of an Object by Rays Passing through a Small Aperture.—This is a necessary consequence of the rectilinear propagation of light. Thus, let a small aperture be made in the shutter C of a darkened room

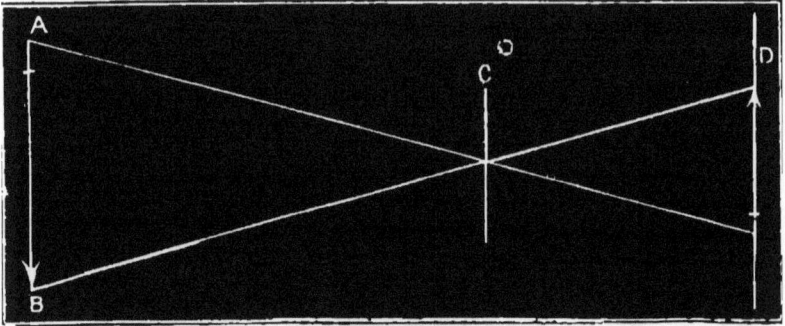

Fig. 18.

(fig. 18), an inverted image of the external object A B will be formed upon the opposite wall at D, *the inversion being caused by the crossing of the rays.* The shape of the image, in such a case, is precisely the same as that of the object, and is independent of the form of the aperture.

30. Shadow—Penumbra.—The shadow which an opaque body casts behind it when exposed to light is

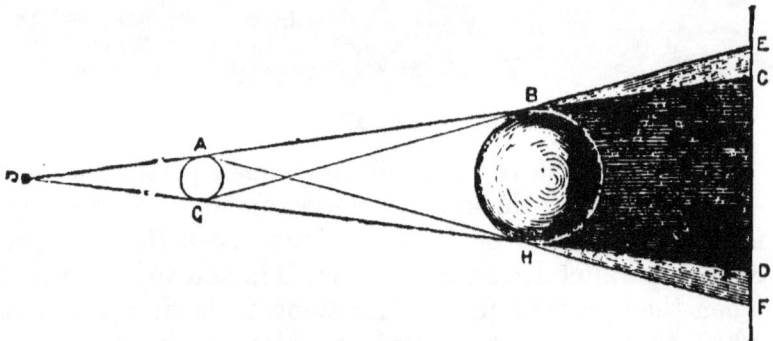

Fig. 19.

also a consequence of the same principle. In the case of a luminous point, it is easy to define or mark off the

8 E   C

shadow. If, however, we have a luminous body besides the true shadow, there is produced also a *partial* shadow or *penumbra*, as it is called. Thus, if P be a luminous point (fig. 19), and B an opaque spherical body, the shadow which it casts behind is marked off by the black part of the diagram, and shows itself on the screen, E F, as a dark circular disc, having a diameter C D. But if A G be the luminous body, the shadow B C D H is fringed with the lighter spaces B C E, H D F, which are exhibited on the screen as a shady circular ring surrounding the disc C D. This constitutes the *penumbra*.

In a total eclipse of the sun, the region of the earth where it occurs is in the shadow cast by the moon; the tract of country, again, from which a partial eclipse is observed, is in the penumbra. Beyond the penumbra there is no eclipse visible.

31. Velocity of Light.—The transmission of light from one point to another is not instantaneous. The rate of transmission, however, is so great, that for any distance on the earth's surface ordinary observation cannot detect any appreciable interval of time between the

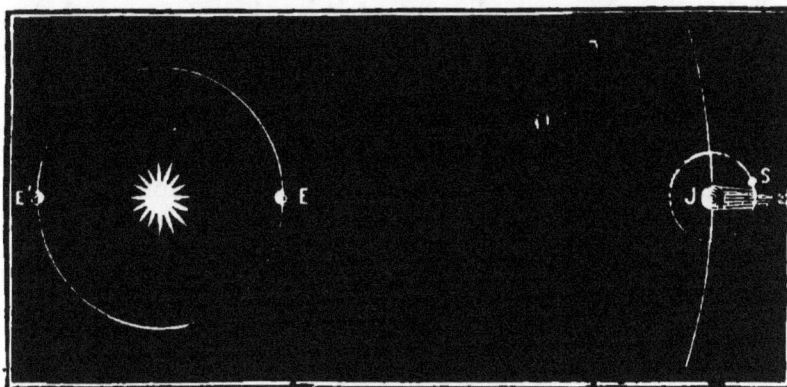

Fig. 20.

occurrence of the luminous phenomenon and its reaching the eye. The velocity of light was first established by Rœmer, a Danish astronomer, in 1675. He deduced it from observations on the first satellite of Jupiter.

Fig. 20 will show the nature of the investigation. By a series of careful observations, he found that the eclipse of the satellite S took place about 15 minutes sooner when the earth was at E nearest to Jupiter, than when at E' farthest away from him. This difference of time he very properly accounted for thus: that the last glimpse of light sent off from the satellite previous to its passing into Jupiter's shadow was delayed this interval in traversing the earth's orbit, that, in fact, the light took 15 minutes to cross from E to E'. In this way Rœmer determined the velocity of light to be 192,500[*] miles per second.

This prodigious velocity will be more readily conceived of, when we state, that whilst light takes about 7 minutes to travel from the sun to the earth, a cannon ball retaining its initial velocity of 1600 feet per second, would perform the same journey in 17 years, and an express train going at the rate of 40 miles an hour in 265 years.

Notwithstanding this enormous speed, the nearest stars are so far off that their light takes between three and four years to reach us; and it has been presumed that the more distant stars in the universe are so remote, that the light from them may take hundreds or even thousands of years to reach our globe.

32. Law of Inverse Squares.—The intensity of light diminishes with the distance from the luminous body according to the same law as that in regard to sound (Art. 9). It is well illustrated by the accompanying diagram (fig. 21).

Let A B, C D, E F, be three surfaces placed respectively at the distances, 1 ft., 2 ft., and 3 ft. from a lamp at L. The same amount of light which is cast upon A B would

* Foucault has devised an ingenious apparatus, by means of which he has determined the velocity to be 185,157 miles per second—a result nearly agreeing with that of Rœmer, if we accept the more generally received opinion among modern astronomers in regard to the real distance of the earth from the sun.

be cast upon C D, a surface four times as great, therefore the intensity of light there is one-fourth of what it is
at A B. Again, the same amount of light which is cast
upon A B would be cast upon E F, a surface nine

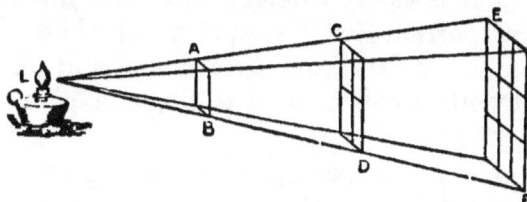

Fig. 21.

times as great, hence the intensity there will be one-
ninth of what it is at A B, and so on; the distances
being, therefore, expressed by the numbers 1, 2, 3, 4, 5,
etc., the intensities will be expressed by the numbers
$1, \frac{1}{4}, \frac{1}{9}, \frac{1}{16}, \frac{1}{25}$, etc. Such is the numerical expression of
the *law of inverse squares.*

33. Measurement of Light — Photometers. — By
means of the law just explained, we can very easily compare one kind of light with another, and express numerically their relative illuminating powers. One of the

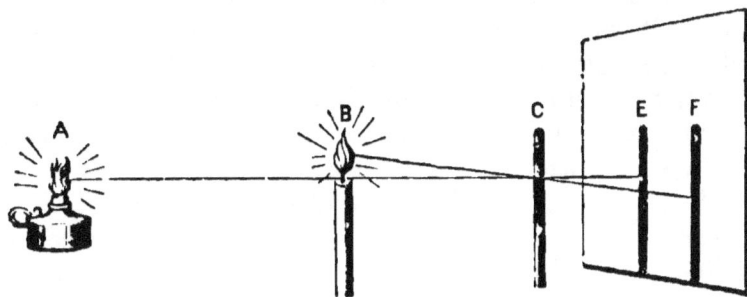

Fig. 22.

simplest methods is what is known as the *"shadow test."*
It consists in making the two lights, A and B (fig. 22),
cast shadows of a rod C upon a screen, and adjusting the lights till the shadows at E and F are illuminated to an equal degree. Now, since the shadow E is

illuminated by the light B, and the shadow F by the light A, and these shadows are equally bright, it follows that the lights A and B cast the same quantities of light on the screen at their respective distances. From the previous law it is easily inferred that the intensities are *in direct proportion* to the squares of these distances. Hence, if B be 2 feet from the screen, and A 5 ft., the relative intensities will be as 4 to 25, or A gives $6\frac{1}{4}$ times as much light as B.

The art of thus comparing or measuring one light with another is called *Photometry*. Several instruments have been constructed with the view of carrying out the same object; these are termed *Photometers*.

CHAPTER II.

34. Reflection of Light—Irregular or Scattered.— Objects are rendered visible to us in consequence of their powers to reflect or scatter the light which falls upon them in all directions. If an object did not possess this power, it would be invisible. It is owing to irregular reflection from the particles of the air and the vesicles of moisture in our atmosphere that we have the sun's light so universally diffused, and gladdening so unsparingly the entire animal and vegetable creation.

35. Light in Itself Invisible.—A beam of light entering by a shutter in a darkened room is rendered visible by its illuminating the particles of dust in its track. Were there no dust particles the beam would be invisible. A striking proof of this is afforded by placing the end of a poker made white-hot at some point in the course of the beam, the dust particles are burnt up, and the end of the poker is shrouded in darkness. If several white-hot bodies be so placed, the beam is seen broken up into several parts.

36. Regular Reflection—Plain Mirrors.—If light fall upon a polished surface, such as a plain mirror, it is *regularly* reflected, that is, it is sent off the reflector in a definite direction. Thus, let A B be a plain mirror (fig. 23). If the ray of light fall perpendicularly upon it, as in the direction F D, it is reflected directly back again. But if it come in the direction C D, then it is reflected in the direction D E, the angle C D F being equal to the angle F D E. The angle C D F is called the *angle of incidence*, and the angle F D E the *angle of reflection*. Moreover, the *incident* ray C D and the *reflected* one D E are in the same plane, which is perpendicular to the reflecting surface. The law of *regular* reflection, therefore, may be expressed thus: *the angle of incidence is always equal to the angle of reflection, the incident and the reflected rays being in the same plane, which is perpendicular to the reflecting surface.*

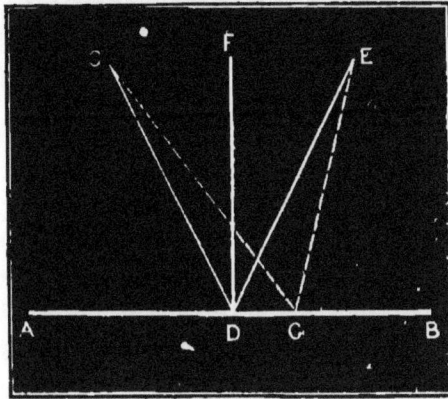

Fig. 23.

It may be proved by geometry that the course C D E is the *shortest* possible from the points C, E, to the mirror. It is shorter, for example, than the course C G E.

37. Influence of Obliquity.—The number of rays which are reflected from a regularly reflecting surface depends upon the magnitude of the angle of incidence. Thus, in the case of water, it has been found that out of 1000 rays, when the angle is 40°, 22 rays are reflected; 60°, 65 rays; 80°, 333 rays; and 89½°, 721 rays.

38. Formation of Images by Plain Mirrors.—When an object is placed before a plain mirror, its image is seen *as far behind the mirror* as the object itself is before it. This is a consequence of the foregoing law.

Firstly, let us take the case merely of a point A (fig. 24); the rays from it, after reflection by the mirror, enter the eye of a spectator at E in a state of divergence— the eye receives these rays as if they came from the point A' behind the mirror. By geometry it is easily proved that the distance A'0 = A0; hence the truth of the proposition.

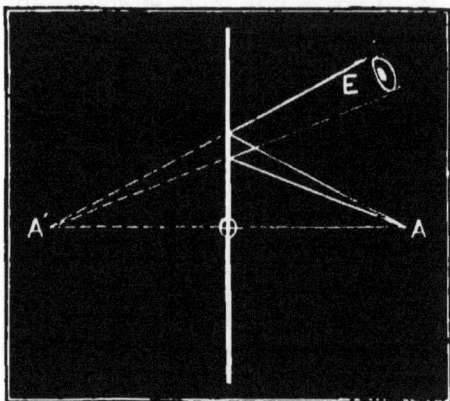

Fig. 24.

Secondly, let A B be the object (fig. 25), the rays from A, after reflection by the mirror, enter the eye as if they came from a real object at A'; similarly, the rays from B enter the eye as if they came from B', and intermediate points in A B are seen at intermediate points in A'B'. Thus, an image of A B is seen at A'B'. The image thus formed is called a *virtual* image.

39. Lateral Inversion.— The image formed by a plain mirror has the same size and shape as the object, but differs in regard to *position*. If, for example, a person stands before a looking-glass, his right eye is the left in the image, and his

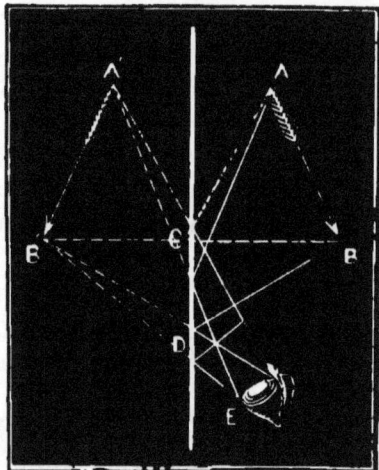

Fig. 25

left eye the right in the image. This effect is known as *lateral inversion*. Hence, writing written backward is adjusted by being held before a mirror, and can be read as if it were written in the ordinary way. A set of types arranged for printing can be read off in like manner. The blocks made for wood-cut illustrations may be examined in this way before they are actually used.

40. Simple Experiments — Curious Facts — Some interesting experiments may be performed with a common looking-glass. If a candle be held directly between the eye and the mirror, one image only of the candle is seen. Now, if the candle be moved gradually towards the side of the mirror, a series of images more and more detached from each other are observed, the second of the series being the brightest and best defined. The experiment is more successful in a darkened room, and with a mirror having a thick glass.

"The first image of the series arises from the reflection of the light from the anterior surface of the glass. The second image, which is usually much the brightest, arises from the reflection at the silvered surface of the glass. . . . . The other images of the series are produced by the reverberation of the light from surface to surface of the glass. At every return from the silvered surface a portion of the light quits the glass and reaches the eye, forming an image; a portion is also sent back to the silvered surface, when it is again reflected. Part of this reflected beam also reaches the eye and yields another image. This process continues; the quantity of light reaching the eye growing gradually less, and as a consequence the successive images growing dimmer, until finally they become too dim to be visible." *

If the angle of incidence be made large by holding the candle very close to the mirror, whilst the eye is placed in a corresponding position, the *first* image may be made to appear as bright or even brighter than the second image, affording a striking proof of the influence of obliquity.

If, when a mirror is set at an angle of 45°, an object be placed vertically before it, its image appears in a

* Tyndall's *Notes on Light*, p. 10.

horizontal position, and if horizontally its image appears vertical.'

If a mirror be moved parallel to itself, either from or towards an object, the image moves twice as fast; or, if it be made to rotate, the angle through which the image moves is twice the angle through which the mirror moves. These facts all follow from the law of reflection, and are not difficult of proof.

If an individual, standing before a large mirror, see his whole person, there is only *half* the length of the mirror concerned in the production of the image. The reason of this will appear from fig. 26. By geometry

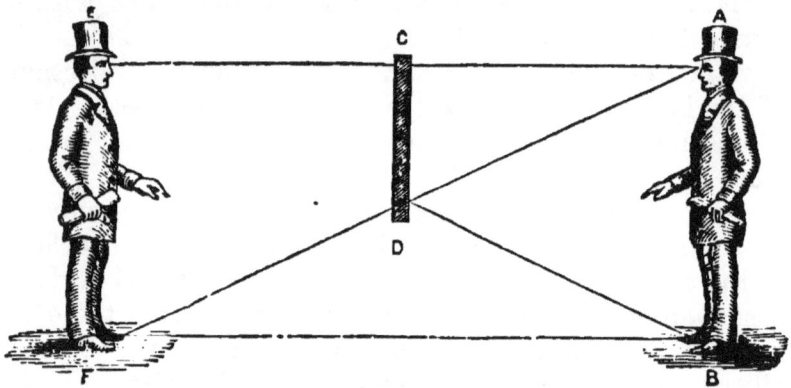

Fig. 26.

the triangles A C D, A E F, are similar, therefore A C bears the same proportion to A E as C D does to E F, but $A C = \frac{1}{2} A E$ (Art. 38), therefore $C D = \frac{1}{2} E F$.

41. Polemoscope.—Light, like sound, is capable of repeated reflection. In the *polemoscope* (fig. 27), an instrument of some service in the time of war, two mirrors are used. They are adjusted at an angle of 45° to the horizon. The upper mirror being directed towards a distant object, the rays of light from the object are reflected by it and sent down upon the lower mirror, when they are again reflected, and where an image of the object is seen.

The officers behind a fortification or parapet can, with

this instrument, watch the movements of the enemy without exposing themselves to danger, and can thus give orders to their men how to direct their fire to the best advantage.

Fig. 27

42. Multiplication of Images—The Kaleidoscope.— When two plane mirrors are set at right angles to each other, an object placed between them yields three images. Thus, let BC, C D (fig. 28), be the mirrors, and A the object. An image of A is formed in the mirror B C at A', a second in the mirror C D at A", whilst a third image is formed by a double reflection of the rays at A'''. The three images and the object are in the angles of a rectangle. If A be at equal distances from the mirrors, they are in the angles of a square. The number of images increases as the angle between the mirrors diminishes. If the angle be 60°, there are 5 images ; 45°, 7 ; 30°, 11. In general, to find the number we have only to divide 360° by the angle between the mirrors, and diminish the quotient (if a whole number) by unity. Hence, if the angle be 0°, that is, if the mirrors be par-

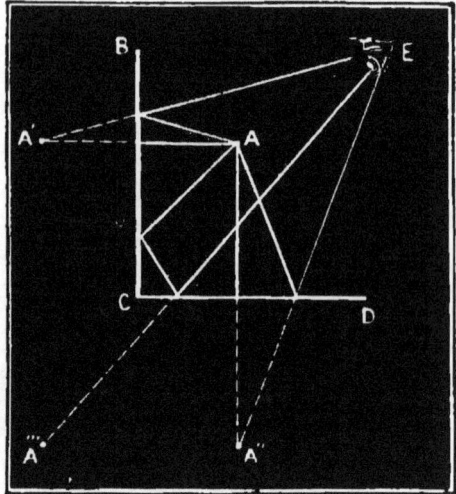

Fig. 23.

allel, the number of images is infinite, but practically the images become in the end so feeble as to cease to be visible. That there is *theoretically* an infinite number of images, may be seen from the following reasoning : Let A and B be the two mirrors, and C an object placed between them. An image of C is formed at C′ by the mirror A, as far behind as the object is before ; but C′ serves as an object for the mirror B, an image of it, therefore, is formed at C″, as far behind B as C′ is before it. Similarly an image of C″ is formed at C‴ by the mirror A, and so on *ad infinitum*.

An arrangement of this kind is sometimes called the "endless gallery," and is used in ball-rooms, picture galleries, jewellers' shops, etc., in order to add to their appearance and produce a dazzling effect.

The *kaleidoscope* (invented by Brewster) depends for its effect upon the multiplication and symmetrical arrangement of images. It consists of a tube of metal or cardboard, in which are placed two strips of silvered glass set

at an angle ; at the end are the small objects, such as
pieces of coloured glass, beads, straws, etc., confined between two glass discs. Looking through the narrow
aperture at the other end, and turning round the instrument, an infinite variety of arrangement is effected in the
small objects, and therefore also an infinite number of
beautiful forms is presented to the eye.

43. Reflection from Curved Mirrors.—The most
common forms of curved reflectors are the *concave*
spherical and the *convex* spherical.

The first of these is the more important to notice. Let
A B be the mirror (fig. 29), O the centre of the spherical

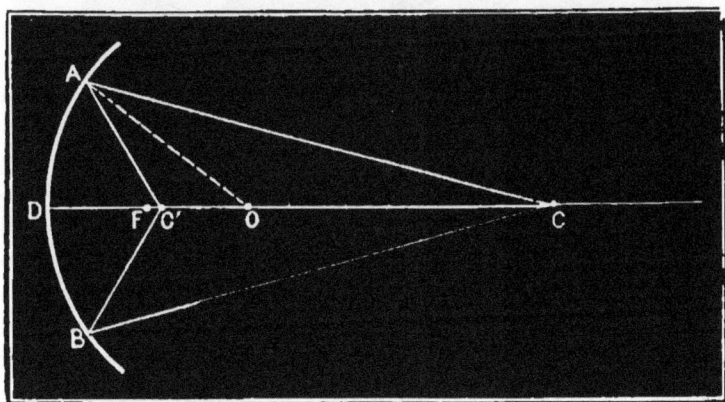

Fig. 29.

shell, of which the mirror forms a portion, C D a line
drawn through O and the middle point of the mirror.
This line is termed the *principal* axis. Rays passing
from the point O are reflected directly back. If the rays
come from an infinite distance, or from the sun, they may
be considered as coming in parallel directions, and
after reflection by the mirror, they are concentrated
in the point F, midway between O and D. This
point is called the *principal* focus. But if they come
from a point C, the divergent beam is concentrated at
some point C' such that the angle C A O is equal to
the angle C' A O, and an image of C is thus formed at
C'. Let now the point C approach the mirror, the focus

C' will move towards O. Passing O, the focus of the
rays moves along O C, until the point comes to C', when
C now becomes the position of the image. The two

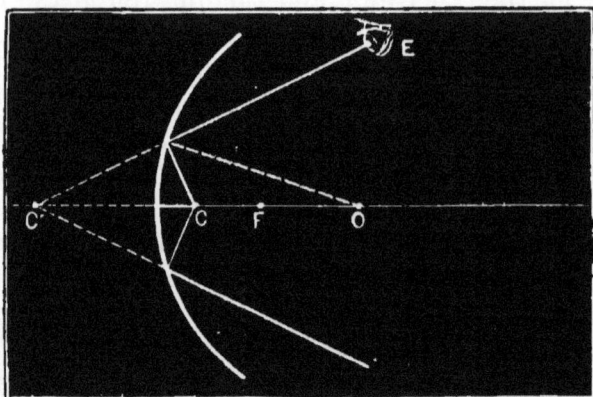

Fig. 30.

points C, C', are thus interchangeable—they are called
conjugate points or foci.   When the luminous point
coincides with F, the rays after reflection pass in parallel directions.   If the point still approach the mirror,
the rays become divergent, and form no *real* focus, but
if produced backwards, as in fig. 30, they meet in some
point C', that is, an eye placed at E will receive the rays
as if they came from C'.   In such a case C' is called a
virtual focus.

If the luminous point be *not* placed on the principal
axis, the position of its image is determined as in
fig. 31.   Draw C D through the point O—this is termed
a *secondary* axis.   The rays are brought to a focus upon
this axis at some point C', between the principal focus
and the centre of curvature, as before.

The formation of the image of an object by this kind of
mirror will now be easily understood.   Let A B be the
object (fig. 32); the rays from A will be brought to a
focus at A', and the rays from B at B'.   Thus there will
be formed between F and O an image A' B', smaller
than the object, and *inverted*.   Similarly, if A' B' be the

object, A B will be its image. Both these images are formed in the air in front of the mirror, and are, therefore, *real* images.

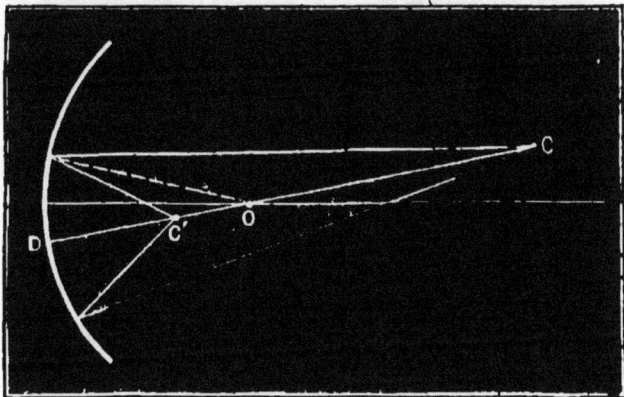

Fig. 31.

If the object be placed between the principal focus and the mirror (fig. 33), then the rays from the object

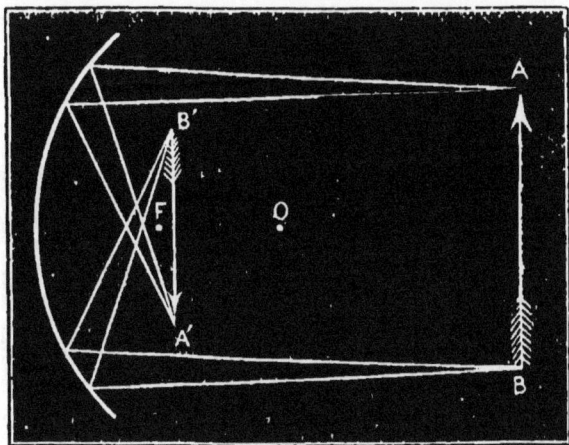

Fig. 32.

A B enter the eye at E, as if they came from an object behind the mirror at A′ B′. In this case the image has the *same* position as the object, and is magnified. It is a *virtual* image.

In regard to the *convex* spherical mirror (fig. 34), let A B be the object, O the centre of curvature, and F the principal focus (evidently *virtual*). The rays from the

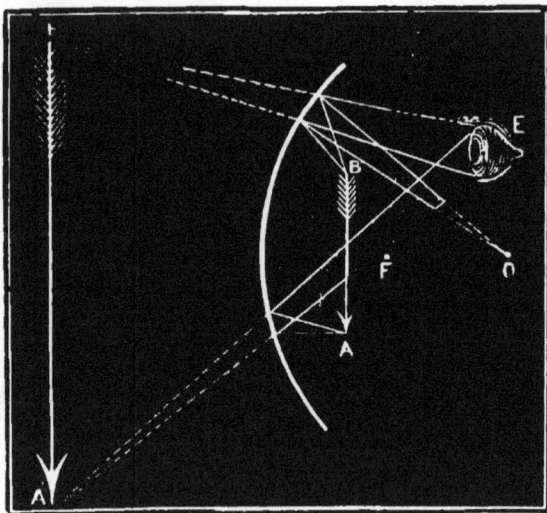

Fig. 33.

object A B, after reflection by the mirror, proceed as if they came from an object at A′ B′. Thus a *virtual* image is seen there smaller than the object, and in the same position.

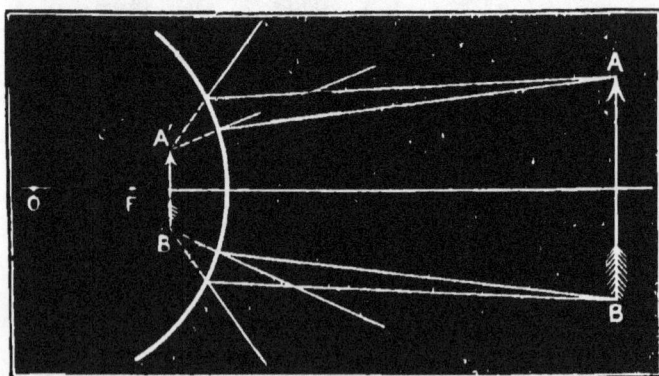

Fig. 34.

44. Spherical Aberration—Caustics.—All the rays

from a luminous point which fall upon a concave spherical reflector, are not concentrated into a *single* point, as we have been supposing. The rays which fall upon the marginal parts of the mirror are not thus concentrated—these, by their intersection with each other, give rise to a series of images forming a luminous surface, which is called a *caustic*. The inability of a concave mirror to collect the rays falling upon it into one point is called *spherical aberration*.

It may be so far obviated by interposing an opaque diaphragm in such a way as to restrict the rays to a small portion of the mirror round the principal axis. The caustic curve may be well seen by placing a common glass tumbler nearly filled with milk beside a candle; the rays are thrown down by the interior face of the glass, and exhibit the curve upon the lacteal surface.

CHAPTER III.

45. Refraction of Light.—A ray of light, in passing from one medium into another, is said to be *refracted* when it deviates from the direction in which it was proceeding before entering the new medium. The deviation itself is called *refraction*. Thus, take the media air and water. Let A O be a ray passing from air into water (fig. 35). If it enter the water in the perpendicular direction A O, it goes straight through without suffering any deviation; but if it enter in the direction A′O, in stead of pursuing the straight course O C, it is bent

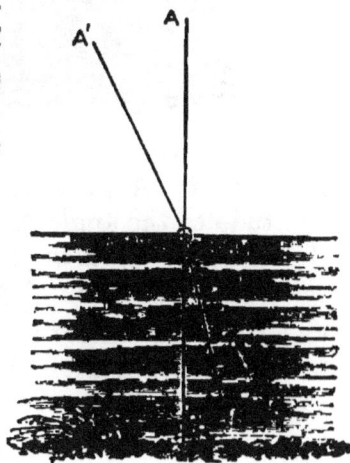

Fig. 35.

from it, and takes a new direction O D. The angle A′O A
is termed the *angle of incidence*, and D O B is termed the
angle of refraction.

It will be seen, therefore, that the behaviour of the
ray is this: when it passes from air into water it is
refracted *towards* the perpendicular, and, conversely,
when it passes from water into air it is refracted *from*
the perpendicular.

46. Law of Refraction.—But the refraction of light
obeys a more definite principle than that just mentioned.
To understand what that is, suppose we describe a circle
with any radius O A (fig. 36), and draw A′ F, D G, per-

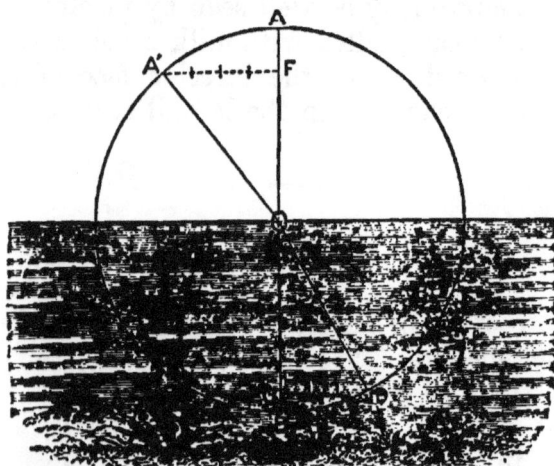

Fig. 36.

pendicular to A B, then it is found whatever be the
magnitude of the angle A′ O A, that the relation between
A′F and D G is always the same, or, as it is often
expressed, the ratio $\frac{A'F}{DG}$ * is a constant quantity.
This constant quantity is called the *index of refraction.*
For air and water the index is $\frac{4}{3}$, and for air and glass $\frac{3}{2}$.
If the course of the ray be reversed, the indices are

* If the radius of the circle be 1, A′ F is the sine of the angle
of incidence, and D G the sine of the angle of refraction; hence,
the law may be expressed thus: "the sine of the angle of incidence bears to the sine of the angle of refraction a constant ratio."

8 E   D

respectively $\frac{3}{4}$ and $\frac{2}{3}$. In general, the greater the refractive index between two media, the greater the deviation of the ray from its original path.

47. **Effects of Refraction.**—The refraction of light explains a number of familiar phenomena. A pool of water appears shallower than it really is. To understand this, let A (fig. 37) be a point in the bottom, the rays A B, A C, in emerging from the water are refracted in the directions B E, C E, and enter the eye there as if they came from the point A′ near the perpendicular, that is, the point A will be seen at A′. The same is true of every other point, hence the whole bottom of the pool appears lifted up. From this it is manifest that

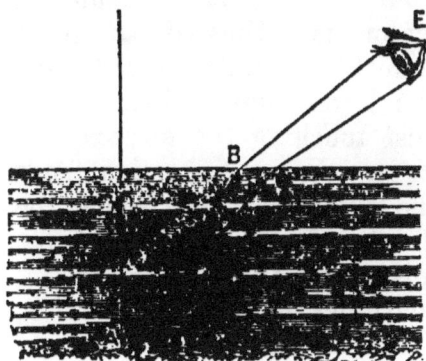 

Fig. 37.   Fig. 38.

the more divergent the rays B E, C E are, in other words, the greater the obliquity of the vision, the shallower will the pool appear.

A stick placed vertically in water (fig. 38) appears shortened, and placed obliquely appears bent, the immersed portion being raised by refraction.

An object under water appears not only less deep, but also of a different shape (fig. 39). Thus the object A B will appear to have the position and shape A′ B′.

A boat floating in clear water seems to have a flatter bottom than it really has, so also a deep-bodied fish seems contracted.

A striking effect of refraction is exhibited by the following experiment:—Place a coin in a bowl, and retire until you just lose sight of the coin by the interposition of the edge. Now desire a companion to fill the bowl with water, the coin again comes into view.

Fig. 39.

In consequence of refraction by the atmosphere, we never see the heavenly bodies in their true places, except those which are directly over our heads. The amount of displacement near the horizon is estimated at about half a degree, but it diminishes rapidly towards the zenith. When we see the lower edge or limb of the sun or moon apparently just touching the horizon, the whole disc is actually below it. Hence refraction tends to prolong the stay of the sun and moon above the horizon—it hastens their rising, and delays their setting.

Even after the sun has disappeared below the horizon, the refraction of his rays continues for some time, which, combined with reflection, produce the phenomenon of *twilight*, by which we pass, with so pleasing a gradation, from the effulgence and activity of day to the darkness and stillness of night.

48. Refraction is always Accompanied by Reflection. —Wherever there is refraction, there is also reflection. We cannot have the one without the other. Should the one disappear, so will the other. The higher the refractive power of a substance, the greater the amount of reflection; hence the striking brilliancy of the diamond.

It will not be difficult, therefore, to understand the appearance presented on the margin of a river or lake. Thus, in fig. 40 the rays from the objects on the opposite bank partly enter the water, suffering refraction, and

are partly reflected from the surface towards the observer.
In virtue of this partial reflection, inverted images of

Fig. 40.

the objects are seen, and they are *feeble,* because of the
loss of the light which passes into the water.

**49. Transparency — Opacity of Transparent Mix-
tures.**—There is no body perfectly transparent, that is,
none which allows perfect freedom in the transmission
of light.   Water, for instance, is transparent at ordinary
depths, but even then a number of rays are quenched.
At the depth of a few hundred feet it would lose all its
transparency.   The dimness of the sun and moon in the
horizon is owing to some of the light being quenched in
its passage through the atmosphere.   Were our atmosphere 700 miles high, we should have no sunlight.

"In the passage from one medium to another of a
different refractive index, light is always reflected; and
this reflection may be so often repeated as to render the
mixture of two transparent substances practically im-
pervious to light.   It is the frequency of the reflections at the limiting surfaces of air and water that
renders *foam* opaque.   The blackest clouds owe their
gloom to this repeated reflection, which diminishes their
transmitted light, hence also their whiteness by *reflected*
light.   To a similar cause is due the whiteness and im-
perviousness of common salt, and of transparent bodies

generally when crushed to powder. The individual particles transmit light freely; but the reflections at their surfaces are so numerous that the light is wasted in echoes before it can reach to any depth in the powder. The whiteness and opacity of writing paper are due mainly to the same cause. It is a web of transparent fibres, not in optical contact, which intercept the light by repeatedly reflecting it. But if the interstices of the fibres be filled by a body of the same refractive index as the fibres themselves, the reflection of the limiting surfaces is destroyed, and the paper is rendered transparent. This is the philosophy of the tracing-paper used by engineers. It is saturated with some kind of oil, the lines of maps and drawings being easily copied *through* *it* afterwards. Water augments the transparency of paper, as it darkens a white towel; but its refractive index is too low to confer on either any high degree of transparency." *

50. **Total Reflection — The Limiting Angle. —** In order that a ray of light may *pass* from a dense medium into a rarer, the angle of incidence must not exceed a certain limit. For water and air this angle is about $48\frac{1}{2}°$, and is called the *limiting* or *critical* angle of refraction. Thus, let A B be the incident ray (fig. 41), then if the angle A B C = $48\frac{1}{2}°$, the refracted ray will emerge in the direction B E, or parallel to the surface of the water. If the angle A B C be greater than $48\frac{1}{2}°$, then the ray is *wholly* reflected, that reflection obeying the ordinary law. It follows from this, that

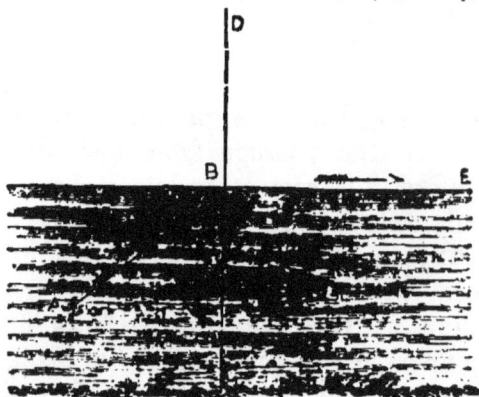

Fig. 41.

all the incident light embraced in the angular space

* Tyndall's *Notes on Light,* p. 19.

D B E, is condensed by refraction into the space A B C, or that the *whole* light which passes into the water is condensed into an angular space of 97°.

We can imagine, therefore, what kind of appearance is presented to a diver, in still shallow water; when he looks upwards, all external objects will be seen, as it were, through a circular aperture overhead of 97° in diameter, whilst beyond this circle he will see, by the effect of total reflection, the various objects at the bottom as distinctly as if he looked directly at them. A man standing on the shore, as well as the shore itself, would appear to be lifted up.

Total reflection may be well illustrated by placing a coin in a tumbler of water, and sloping the tumbler till the light acquires the proper incidence. On looking upwards a distinct image of the coin is seen towards the surface of the water. In an aquarium, if the eye be directed to the surface of the water, the various objects in it may be rendered visible in a like manner.

51. Lenses—Converging and Diverging.—A lens is a portion of a refracting substance, such as glass, having its bounding surfaces either both curved, or the one plane and the other curved. Lenses are of two classes, *converging* and *diverging*, and are named from the form of their external surfaces. Each class comprises three kinds.

Converging.   Fig. 42.   *Diverging.*

Thus (fig. 42), A is called a *double convex* lens; B, a *plano-convex;* C, a *concavo-convex* (or *meniscus*), the *convex* surface having the greater curvature. D is called a *double concave* lens; E, a *plano-concave*, and F, a *convexo-concave*, the *concave* surface having the greater curvature.

The effect of a *converging* lens as A, and of a *diverging* lens as B, on a beam of light, will be understood from figs. 43, 44.

Let the beam consist of parallel rays; the lens A (fig. 43) brings the rays to a focus at the point F. This point is called the *principal* focus, that is, it is the focus of parallel rays. It is a *real* focus.

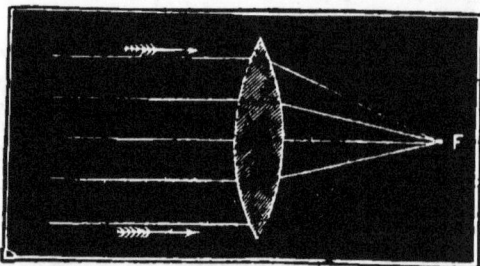

Fig. 43.

Again, the lens B (fig. 44) causes the rays to diverge, as if they came from a point F', on the same side of the lens on which the light falls. This point is therefore the *principal* focus. It is evidently a *virtual* focus.

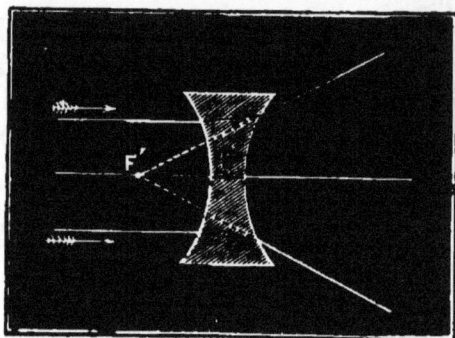

Fig. 44.

It may be noticed that a converging lens is thicker and a diverging lens thinner at the centre than at the exterior borders. They may therefore be distinguished very easily in this way.

52. Formation of an Image by a Double Convex Lens.—Let us first take the case of a luminous point placed before a double convex lens. Let A be the luminous point (fig. 45); draw A A' through the centre of the lens, perpendicular to its two convex surfaces—this is termed the *principal* axis. The rays from A are brought to a focus at a point A' beyond the principal focus F, and a *real* image of A is formed there. These two points, as before, are convertible; they are conjugate foci. If A now be moved towards the lens,

A' will retire from it, until A coincides with the principal focus F' (O F being equal to O F'), when the rays, as in fig. 43, will emerge from the lens in parallel directions.

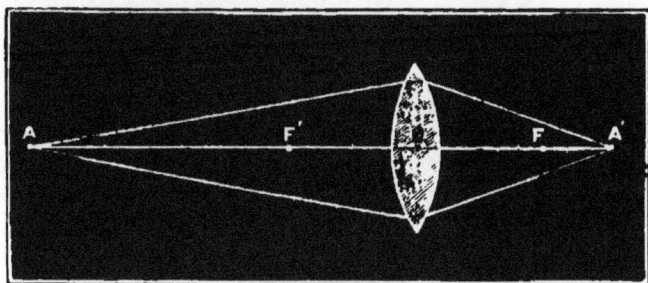

Fig. 45.

If A be placed between F' and the lens (fig. 46), the rays, after passing through the lens, are divergent, proceeding as if they came from a point A'. The point A' is therefore a *virtual* focus.

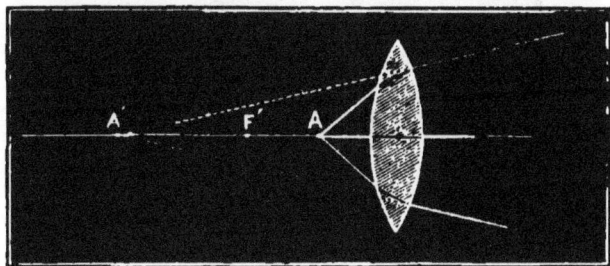

Fig. 46.

Let now an object A B be placed before the lens beyond the principal focus (fig. 47). Draw A O A' and B O B' through the centre of the lens. The rays from A are brought to a focus at A', those from B at B', and the rays from intermediate points in A B at intermediate points in A' B'. Thus a real inverted image of A B will be formed at A' B'. This image may be seen by an eye placed beyond A' B', or it may be projected on a screen, whose distance from the lens is equal to that of A' B'. The size of the image bears the same proportion

to the size of the object, as the distance O A' does
to O A.

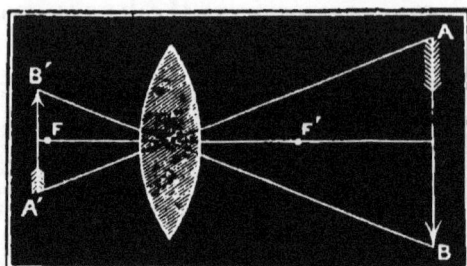

Fig. 47.

If the object be placed *between* the lens and the
principal focus (fig. 48), a magnified and erect image will

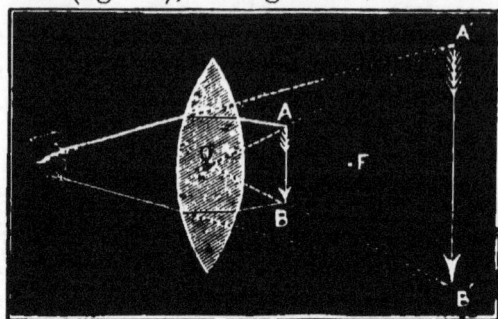

Fig. 48

be formed. Then, of course, the image is virtual. Such
an arrangement constitutes the *simple* microscope.

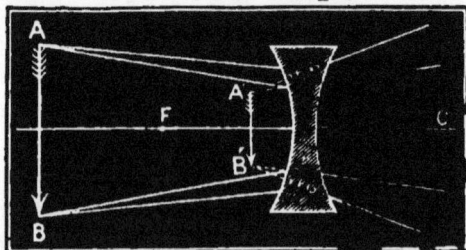

Fig. 49.

**53. Formation of an Image by a Double Concave
Lens.**—We have seen that a double convex lens may give
either a real or a virtual image, according to the distance

of the object. A double concave lens gives only a virtual image at all distances. Let A B be the object (fig. 48), F C the principal axis, F the principal focus. The rays from A B, after traversing the lens, are divergent, and enter the eye as if they came from a real object at A' B', that is, there will be an image of A B seen at A' B', between the lens and the principal focus. That image is erect and smaller than the object.

54. Camera Obscura.—This instrument is represented sectionally in fig. 50. It consists of a sloping, wooden

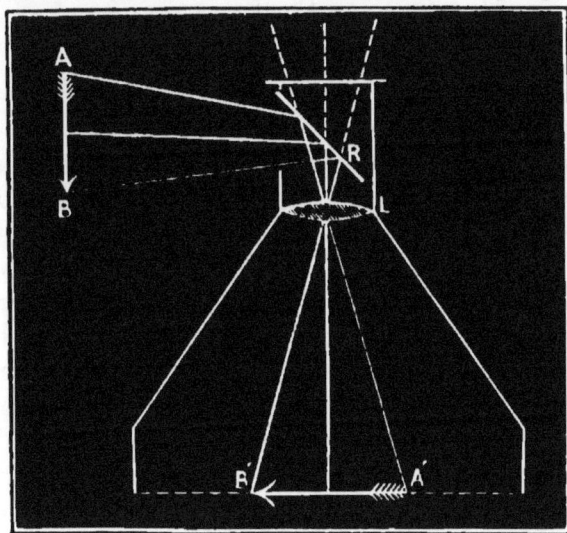

Fig. 50.

box, blackened inside, to obviate irregular reflection, at the bottom of which is placed a sheet of paper. At the top there is a small cubical box, formed of two parts, one of which slides into the other, for the purpose of focal adjustment, and which contain respectively a plane reflector and a double convex lens. The action of the instrument is this: The rays from some distant object, A B, are reflected by the mirror towards the lens; the lens concentrates these rays into a focus, and an image A' B' is thus formed upon the paper. There are two apertures, through one of which the picture is viewed, and through the other the hand may be thrust for the purpose of sketching it off.

55. Magic Lantern.—The construction and principle of this instrument will be understood from fig. 51. The

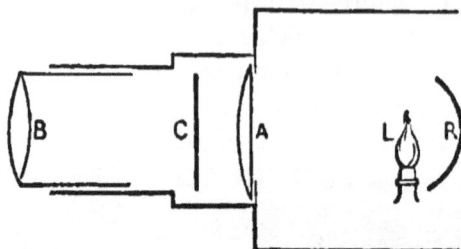

Fig. 51.

lamp L is placed in the focus of a concave reflector R; the rays from it, after reflection, are condensed by the lens A upon the glass slide C, on which the picture is painted. An image of the illuminated picture is then formed by the lens B, and thrown upon the screen in a darkened room. The lens B is fixed in a tube which slides into the other, and its focus can therefore be adjusted to different distances. From what we have seen in regard to the formation of an image by a lens, the slide C must, of course, be introduced in an inverted position. The image on the screen is magnified as many times as the distance of the screen from B contains the distance of C from B.

Dissolving views are produced by having two similar magic lanterns, placed side by side, and directed towards the same part of the screen. A metallic diaphragm is placed before the lanterns, and is so arranged as gradually to close the aperture of the one lantern, whilst that of the other is being opened. By this artifice a pleasing variety of effect is obtained.

56. Spherical Aberration.—We have been proceeding upon the supposition that *all* the light passing through a lens is brought to the same focus. This, in reality, is not the case. The rays which fall upon the exterior borders of the lens are *not* concentrated into the same

point, but are found to intersect each other at different
points, forming a luminous surface, which is called a *caustic*,
by refraction (Art. 44). This inability on the part of a
lens to bring all the rays to a single focus, is called
spherical aberration.

This aberration interferes with the sharpness or distinctness of an image, but may be partly obviated by
interposing an opaque diaphragm provided with a central
aperture. This allows the rays *only* which fall upon
the central part of the lens, to pass through. Recourse
is had to this device in photography.

CHAPTER IV.

57. The Eye: its Structure.—The different parts of
this wonderful organ are exhibited in fig. 52.

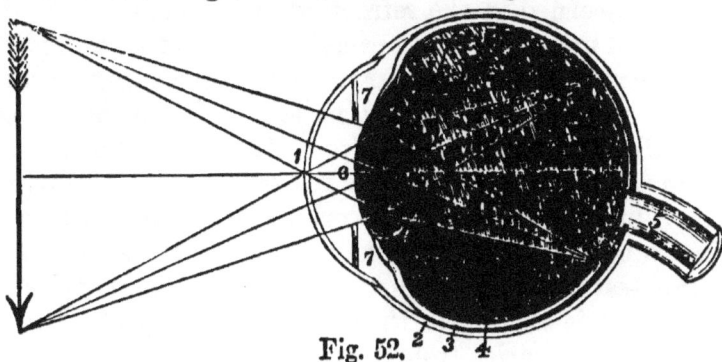

Fig. 52.

1, is called the *cornea;* 2—the *sclerotic* coat, or white of
the eye; 3—the *choroid* coat, covered with a black pigment
to prevent internal reflection; 4—the delicate network
of the *retina*, which, being an extension of the optic
nerve (5), conveys the impression of the image there depicted to the brain. Behind the cornea is an opening
called the *pupil* (6), surrounded by the membrane of the
iris (7), which is differently coloured in individuals,
giving rise to difference of colour in eyes. The iris per-

forms the important function of regulating the quantity of light which passes into the interior chamber of the eye, by its involuntary action in enlarging or contracting the diameter of the pupil. 8, is the *crystalline lens*, more convex behind than before, and consisting of concentric layers of tissue, which increase in consistency towards the centre. The space between the cornea and the lens is filled with a fluid like water, called therefore the *aqueous humour;* whilst the whole of the posterior chamber is filled with another fluid, called the *vitreous humour*, from its resemblance to melted glass.

58. Distinct Vision.—In order that we may see any external object *distinctly*, an image of that object must be thrown upon the retina; in other words, the rays of light from the object must be brought to a focus there, (fig. 52). This is effected chiefly by the intervention of the cornea; but the other parts of the eye, the aqueous humour, the crystalline lens, and the vitreous humour, are all concerned in the refraction of the rays.

That an image of an external object is actually depicted upon the retina, has been shown by experimenting with the eye of a recently slaughtered bullock. What holds good of a bullock's eye, is believed to be true of the human eye. The image also is *inverted*, the reason of which is obvious. For ordinary eyes there is a certain distance at which an object must be placed, in order that it may be seen with the greatest possible distinctness. This *distance of distinct vision*, as it is termed, in the case of small objects, such as common type, varies from 10 to 12 inches.

59. Punctum Cœcum — Foramen Centrale.—It is a remarkable fact, and one which can scarcely be credited, that, though the optic nerve is the medium of communication with the brain, when the image of an object falls upon the *base* of that nerve (fig. 52), there is no impression produced—it is quite insensible to the action of light. The following interesting experiment may be made in corroboration :—Put three spots of ink on a sheet of paper, about three inches apart. Shut one eye, and look

steadily with the other at any of the spots. If now the
head be slowly moved towards either side, up or down, a
position will be obtained where one of the three spots
entirely disappears. By a little care any one of the spots
may be made to vanish. This is owing to the image falling upon the surface in question. From this circumstance
the surface has been called the *punctum cæcum,* or "blind
spot."

Every part of the retina is not equally sensitive to the
action of light. There is a small portion where the
organization seems to be more delicate than any other.
This part is called the *foramen centrale,* or "central opening." * If the image of an object fall upon this, it is seen
with the greatest possible distinctness.

The retina of another person's eye may be examined,
and this part thus rendered visible by an instrument
called the *ophthalmoscope.*

60. Why Objects are Seen Erect.—Since the image of
any external object depicted upon the retina is *inverted,*
a natural question arises, how do we correct this
inversion ? Some have supposed that we actually
do see everything inverted, but that, from habit
and experience, we learn to assign to every object its
true position. According to this opinion, infants see
objects upside down; and it is only by comparing the
erroneous information acquired by vision, with the more
accurate information acquired by touch, that they learn
to see objects as they really are. Others again account
for it by supposing that we really judge of the position
of an object from the *direction in which the rays of light
proceeding from it enter our eyes.*

61. Single Vision.—As there is an image of the
object in each eye, it may be asked, why is it we do not
see *double* when we use both eyes? This question is not
difficult to answer. When we fix our eyes upon an

---

* This name is apt to lead to misconception. There is really
no part of the retina where there is an "opening," except at the
base of the optic nerve.

object, each eye arranges itself in a particular manner. Thus, let B, C be the two eyes (fig. 53), and A the object. Draw Aa, Aa through the centre of the crystalline lens,

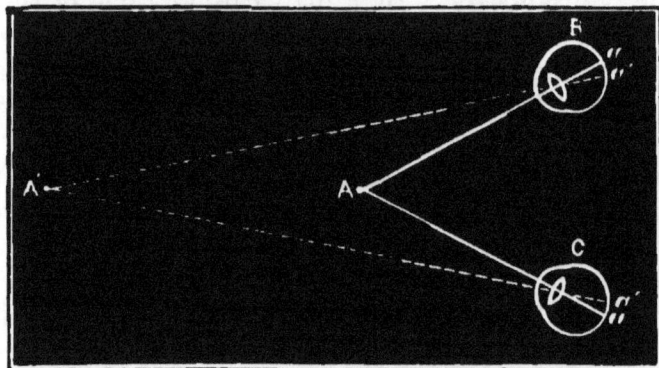

Fig. 53.

and at right angles to the convex surfaces. These lines are called the *optic axes*, and the angle between them, aAa, the *optical angle*. The eyes adjust themselves so that the optic axes intersect each other *at* the object. In consequence of this, a precisely similar image of the object is formed in each eye, and therefore a precisely similar impression of the object is conveyed to the mind. If either eye be prevented from thus adjusting itself by slight pressure on the eye-ball, double vision results. Hence persons who *squint* have always double vision. It thus appears that single vision arises from the circumstance that the image is cast upon *corresponding* parts of the retina in both eyes.

If, whilst the eyes are directed upon a small object at A (fig. 53), there is another object A' placed beyond, that latter object will be seen double. This results from the images in the two eyes being thrown upon *different* parts of the retina. Thus, the image in B is formed on the *left* of the optic axis, and that in C on the *right*. If the eyes be directed upon A', then A will be seen double for the same reason.

62. Adjustment of the Eye for Different Distances.

—Experience teaches us that objects are seen with sufficient distinctness, though their distances may vary considerably. It follows, therefore, that the eye must have
the power of accommodating itself to the distance at
which an object is situated. Several opinions have been
entertained on this matter. Some physicists attribute
it to a property which the crystalline lens possesses of
changing its curvature, so as, in every case, to make the
rays converge to a focus upon the retina. Others suppose that it is dependent upon the contraction and
dilatation of the pupil. According to this hypothesis,
objects at some distances are seen by virtue of the rays
which fall upon the exterior borders of the lens; whilst
again, near objects are seen by virtue of the rays which
pass through the middle of the lens, such rays under-
going thereby a greater degree of refraction than in the
other case. Others, again, imagine that the focal distance of the lens, for different distances, may vary so little
as not to cause any appreciable effect on the distinctness
of the image.

63. Long and Short Sight—Spectacles.—In advanced
life the eye loses its power, and becomes incompetent to
bring the rays to a focus upon the retina at the ordinary
distance of distinct vision. This condition of the eye,
which is common to most elderly persons, is termed "far-
sightedness."

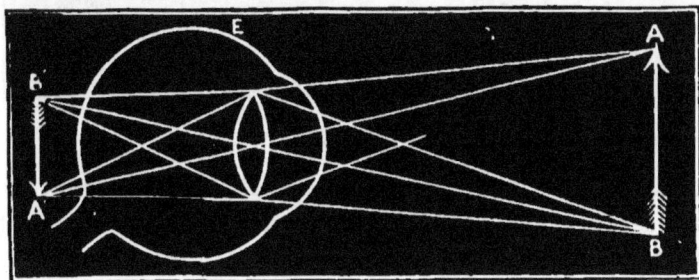

Fig. 54.

Let E be such an eye, and A B the object (fig. 54);
the rays from it tend to converge to a focus *beyond*

the retina, and an image of the object would be formed at A' B'; the rays, therefore, which really fall upon the retina, are in a *state of separation*, each produces a picture of its own, and indistinct vision is the consequence. The defect may be so far remedied by placing the object at a greater distance from the eye, so as to give the rays a less degree of divergence, and thus enable the eye to bring them to a focus upon the retina. Hence, old persons have a greater difficulty in seeing distinctly near objects than those at a distance. But, should the eye be too weak even to accomplish this, a convex lens or glass must be used, just of sufficient power to aid the eye towards the proper convergence of the rays.

Some eyes again have too much convergent power, that is, they bring the rays to a focus in *front* of the retina. This condition of the eye is called "short-sightedness." Thus, if A B be the object placed before an eye of this kind (fig. 55), an image of A B is formed in the interior of the eye at A' B'; the rays, therefore, in this case also, fall upon the retina in a scattered state, and indistinct vision ensues. If the object be placed

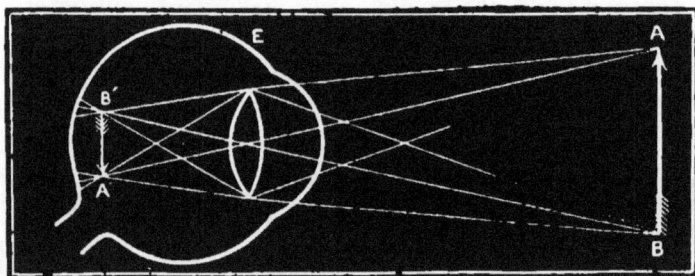

Fig. 55.

nearer the eye, the divergence of the rays is increased, and may be made such as just to enable the eye to form the image upon the retina. Hence short-sighted persons can see near objects with greater distinctness than distant ones.

The remedy for short-sightedness is to provide a concave glass, of such diverging power as to give the eye

8 E   E

sufficient work to do to converge the rays to a focus on the retina.

The particular adaptation of *spectacles* to aid in vision will thus be apparent.

64. Size of Objects—Visual Angle.—The size of the image of any external object depicted on the retina, depends upon the distance of the object from the eye. Thus, let R represent the retina, and L the crystalline lens, A B the object (fig. 56). The size of the image for that distance of the object is *a b*. If, now, the same object be placed at A′ B′, its image becomes reduced to *a′b′*,

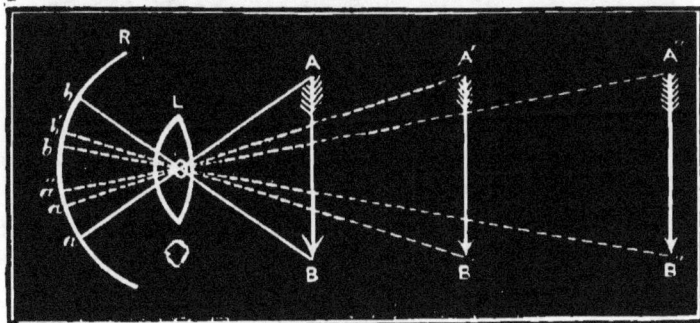

Fig. 56.

and if placed farther away still at A″B″, to *a″b″*. In a word, the greater the distance of the object the smaller the image. The angle A O B is called the *visual* angle; in general, it is *the angle which the object subtends at the centre of the crystalline lens*. It thus appears that, *so far as the eye is concerned*, the size of an object depends upon the magnitude of the visual angle.

If, therefore, we have any number of objects, A, B, C, etc. (fig. 57), having the *same* visual angle, these, though in reality very different in magnitude, will cast the same size of image on the retina.

It thus appears that, were we to judge of the *size* of an object from the size of the picture formed on the retina, we would judge erroneously. How then, it may be asked, can we form so correct a judgment of the size of objects? The reason is, that we learn by habit and

experience to take into account the *distance* at which the
object may be placed.   A child, for instance, placed near

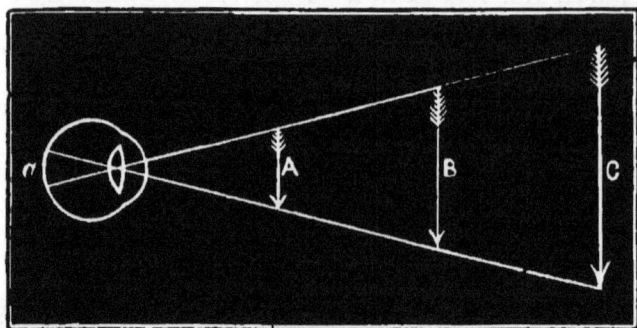

Fig. 57.

us may appear under the same visual angle as a man at
some distance off, yet we are in no way misled as to their
comparative sizes; we do not imagine that the child is as
tall as the man.   We learn by experience to combine,
in our judgment, the *distance* at which the child is in
reference to the man; and thus it is that we are led to
correct the impressions which our eyes of themselves
would convey.

65. **Persistence of Impressions.**—The impression
which light makes on the eye is not obliterated *instantaneously;* it continues for a short time after the cause of
that impression has ceased to act.   Its duration is found
to vary with different eyes, and also with the intensity
and colour of the light; but, in all cases, its amount is a
sensible fraction of a second. If, therefore, a series of distinct impressions be made upon the eye, which succeed each
other with sufficient rapidity, these impressions will be
blended together and will produce a continuous sensation.
This persistence of impression explains the following
familiar facts: The glowing end of a stick which has been
thrust into the fire, when whirled rapidly round, gives
the appearance of a continuous circle of light.   A flash
of lightning is seen for a time as an unbroken track of
fire in the heavens.   A falling star presents a similar

appearance. So also, when it is raining heavy, there appear so many lines of water falling to the ground.

On this principle a number of entertaining instruments have been constructed. The *magic disc*, the *thaumatrope*, the *kaleidophone*, the *wheel of life*, the *chromotrope top*, etc., all owe their action to this principle.

66. Irradiation.—This is the phenomenon in virtue of which small objects, when highly illuminated, appear larger than they really are. It results from the spherical aberration of the eye, or from the fact that there is an extension more or less of the image upon the retina beyond its *true* or *defined* outline.

Irradiation explains such facts as the following:—

"A platinum wire, raised to whiteness by a voltaic current, has its apparent diameter enormously increased. The crescent moon seems to belong to a larger sphere than the dimmer mass of the satellite which it partially clasps. . . . The white-hot particles of carbon in flame describe lines of light because of their rapid, upward motion. These lines are *widened* to the eye; and thus far greater apparent solidity is imparted to the flame than in reality belongs to it." * So also a bright star, such as Sirius or the Dog-star, appears larger than it really is.

67. Stereoscope.—In looking at any object, the image or picture formed in each eye is *not* the same. For example, if we place a vase before us, there is depicted on the retina of the *right* eye an image of the vase, and on that of the *left* eye also an image of the vase; but the former image is *different* from the latter—a part of the vase is seen by the right eye which is not seen by the left, and a part is seen by the left which is not seen by the right. If, therefore, pictures be taken of the vase corresponding to the views of the individual eyes, these pictures will not be identical. The object of a stereoscope is to *combine* such pictures, and thus by its use there is produced in the mind the same impression as would result were the object actually before us. Fig. 58 will

* Tyndall's *Notes on Light*, pp. 26, 27.

show the difference which subsists between the two pictures in a stereoscopic slide.

The first form of the stereoscope is due to Wheatstone. It consisted of two plane mirrors, so arranged as to reflect to each eye the particular view of the object which belonged to it, and at the same time to make these views coalesce. This is known as the "reflecting" stereoscope.

The most·familiar form of the instrument is the "len-

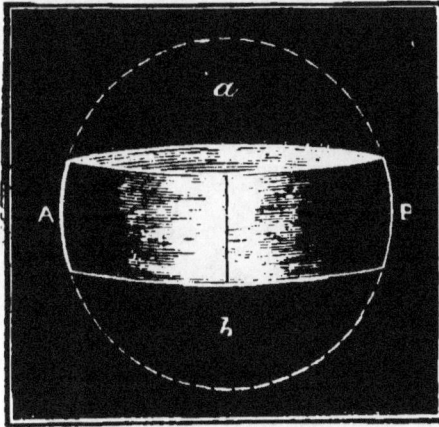

Fig. 59.

ticular" stereoscope, invented by Brewster. Its construction and action will be understood from the accompanying figures. A double convex lens (fig. 59) has its sides, *a*, *b*, cut away, the remaining part A B is then cut across at the middle. The two halves are set in the instrument with their edges A, B in juxtaposition, as in fig. 60.

Now let C, C' be the two pictures of the object placed in the focus of the divided lens, the rays, after emerging from the glasses, enter the eyes as if they came from one picture at D; in other words, the two pictures will overlap or be blended together at that point, and thus

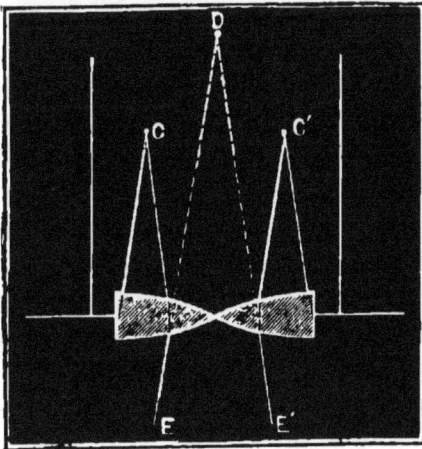

Fig. 60.

there is produced in the mind the impression of *solidity* or *relief*.

CHAPTER V.

68. Medium with Parallel Surfaces.—When a ray of light passes obliquely through a plate of glass with parallel surfaces, it emerges in a direction parallel to the incident ray. Thus, let A B be the plate (fig. 61), the ray C D in entering the glass is refracted in the direction D E, and emerges in the direction E F, E F being parallel to C D. A similar treatment of light takes place with hollow glass vessels having such surfaces, and containing liquid.

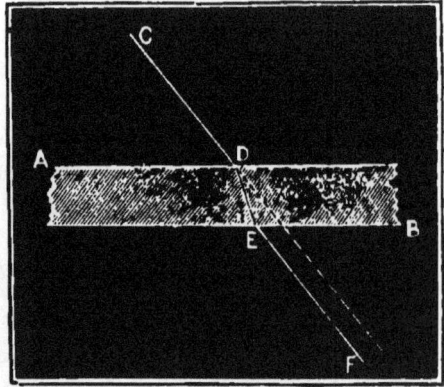

Fig. 61.

69. Prisms—Course of a Ray through a Prism.— A *prism* in optics is a wedge-shaped transparent substance, constructed generally of glass. The angle enclosed by the two oblique faces is called the *refracting* angle of the prism (fig. 62).

The treatment of a ray of homogeneous light by a prism is this : Let S I be the ray (*a*), and I N the perpendicular upon the face at the point of incidence; the ray is refracted *towards* the perpendicular, and follows the course I E inside the prism. On emergence it is again refracted, but now *from* the perpendicular upon the other face, E N', in the direction E B. Thus the ray is bent twice in the same direction, that is, *towards the base of the prism*. If the incident ray (*b*) be perpendicular to the face of the prism, there is only one refraction, and that takes place at the point of emergence, in the direction E R.

If, again, the incident ray (*c*) so fall as that the refracted ray I E becomes parallel to the base, then the emergent ray E R is such that the angle R E N' = the angle S I N.

In this case the deviation of the incident ray from its original course is the *least possible*.

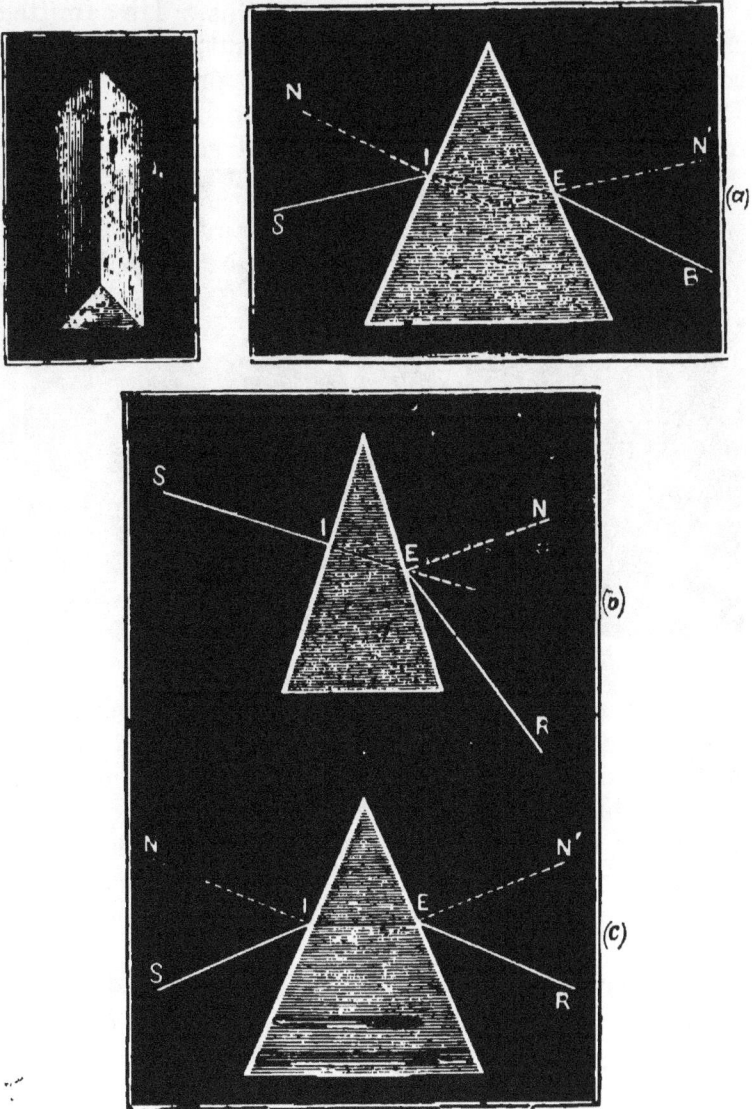

Fig. 62.

70. Dispersion.—When a beam of solar light is made to pass through a prism, the beam is not only refracted, but it is also decomposed or broken up into so many constituent parts, a phenomenon which is called *dispersion*.

Newton was the first to discover this. He admitted a sun-beam, S, by an aperture in the shutter of a darkened room, and allowed it to fall upon a prism P (fig. 63). Placing a screen, E, at some distance, he found an elongated image of the sun there formed, and coloured after the

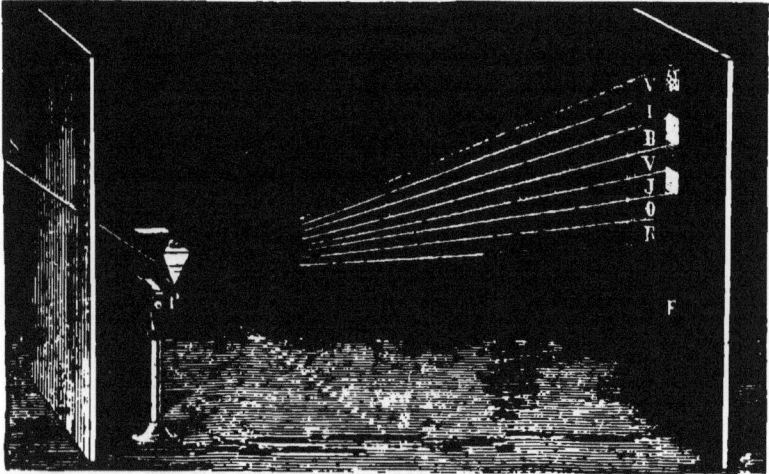

Fig. 63.

following manner (commencing from the lower end):— Red, orange, yellow, green, blue, indigo, and violet. In order to see whether there was any further decomposition possible, Newton transmitted each of these colours separately through another prism, but no other variety of colour was obtainable. The red light gave a red image, the orange an orange image, and so on successively.

From such experiments it has been inferred that the sun's light is not homogeneous, but consists of *these seven different kinds of light,* and these *only.*

The elongated coloured image thus formed by a prism is termed the *solar spectrum.* The different colours have different amounts of refrangibility. Thus the red light is *least* refrangible, and therefore takes the lowest part of the spectrum ; the violet light, again, is the *most* refrangible, and takes the highest part. Hence arises the phenomenon of dispersion. Moreover, the colours do not occupy an equal space in the spectrum. Orange is found to occupy the least space, and violet the greatest.

71. Curious Facts as to the Solar Spectrum.—Beyond the limits of the *visible* spectrum, in both directions, there are rays which do not excite the optic nerve, but the existence of which, though they are invisible, is proved by experiment. There are rays, beyond the red, which have great *calorific* or *heating* power ; and there are rays, beyond the violet, which possess considerable *chemical* power.

The coloured spaces in the spectrum are *not continuous.* There are certain interruptions in their continuity, in the shape of a number of thin dark lines, which are distributed irregularly throughout, in a direction perpendicular to its length. These lines were first carefully studied and accurately mapped out by Fraunhofer, and hence are called *Fraunhofer's lines.*

72. Recomposition of White Light.—Since white light can be broken up into seven different colours, it may naturally be asked, can these colours be so *recombined* as to produce white light ? Yes ; they can. There are several ways in which this may be effected. The following may be mentioned :—

(1) By taking another prism of the same refracting angle as the dispersing one, and placing it near the other in an *inverted* position. The first prism decomposes the solar beam ; the second reunites the constituent parts of it, and produces a *white* image of the sun.

(2) By allowing the decomposed beam to fall upon a concave mirror. The coloured rays after reflection are concentrated in the focus of the mirror, and form there a *white* image, which may be received upon a screen.

(3) By means of Newton's disc. This consists of a disc of cardboard (fig. 64), coloured with the several tints, the different sectors being made to correspond, as far as possible, with the proportional spaces of the colours

as they exist in the spectrum. If this disc be made to rotate rapidly, the colours are so blended as approximately to produce *whiteness*.

Fig. 64.

73. Doctrine of Colours.—" Natural bodies possess the power of extinguishing, or, as it is called, *absorbing* the light that enters them. This power of absorption is *selective*, and hence, for the most part, arise the phenomena of *colour*.

"When the light which enters a body is *wholly* absorbed the body is black ; a body which absorbs all the waves equally, but not totally, is grey ; while a body which absorbs the various waves unequally is *coloured*. Colour is due to the extinction of certain constituents of the white light within the body, the remaining constituents, which return to the eye, imparting to the body its colour.

" It is to be borne in mind that bodies of all colours,

illuminated by white light, reflect white light from *their exterior surfaces.* It is the light which has plunged to a certain depth within the body, which has been *sifted* there by elective absorption, and then discharged from the body by interior reflection, that, in general, gives the the body its colour. . . .

" A body placed in a light which it is incompetent to transmit appears black, however intense may be the illumination. Thus, a stick of red sealing wax placed in the vivid green of the spectrum is perfectly black. A bright red solution similarly placed cannot be distinguished from black ink ; and red cloth, on which the spectrum is permitted to fall, shows its colour vividly when the red light falls upon it, but appears black beyond this position. . . .

" Colour is to light what pitch is to sound. The pitch of a note depends solely on the number of aërial waves which strike the ear in a second. The colour of light depends on the number of etherial waves which strikes the eye in a second. . . .

" The waves of the extreme violet are about half the length of those of the extreme red, and they strike the retina with *double* the rapidity of the red. While, therefore, the *musical scale,* or the range of the ear, is known to embrace nearly eleven octaves, the *optical scale,* or range of the eye, is comprised within a single octave."*

74. Complementary Colours.—One colour is said to be *complementary* to another, when in combination with that other it produces white light. Thus red is complementary to the colour resulting from the mixture of the remaining constituents of the spectrum, or to greenish blue ; yellow, to indigo blue, and so on.

75. Chromatic Aberration.—When white light passes through an ordinary glass lens, there is also a certain amount of dispersion—the lens is incompetent to bring the differently coloured rays to a common focus, in consequence of which the image of an object is seen with a

* Tyndall's *Notes on Light,* pp. 35, 37, 38.

coloured border. This lack of power on the part of a lens is called *chromatic aberration.*

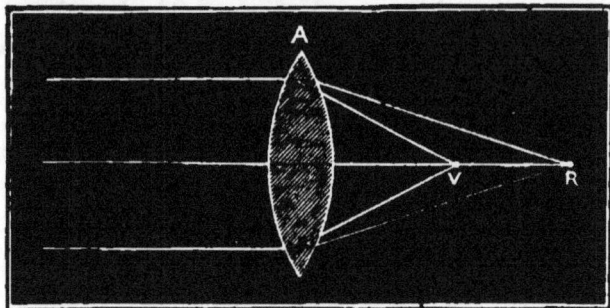

Fig. 65.

Thus the lens A (fig. 65) will decompose the light, and form a *series* of foci instead of one—the red rays being concentrated at R, and the violet ones at V, whilst the intermediate rays are arranged in order.

This defect in a lens is obviated by the combination of a double convex lens of *crown* glass, with a concavo-convex of *flint* glass. The effect of the second lens is to re-blend the coloured rays which the first has produced, and at the same time such an amount of refraction is preserved as to bring the light to a focus.

Such a lens is called an *achromatic lens.* It is much used in the more perfect optical instruments.

QUESTIONS.

1. Explain the *inversion* of the image of an object by rays passing through a small aperture; and why the shape of the image is independent of the *form* of the aperture.

2. It is found that a lamp and a candle, when placed respectively at a distance of 5 feet and 3 feet, illuminate a screen equally. Express the relative intensities of the two sources of light.

3. Explain by a diagram what is meant by the principal axis of a concave spherical mirror. Show also the position of the principal focus.

4. If a ray of light pass from air into water, show by a diagram what course it pursues, and explain clearly the expression the *index* of refraction.

5. Sketch (1) a converging, and (2) a diverging lens. In using the simple microscope, where must the object be placed that a magnified image of it may be seen? Is the image real or virtual?

6. Point out the causes and remedies of long and short sight.

7. Describe and explain the stereoscope.

8. In looking at any object with both eyes, how do they arrange themselves? If I hold up my finger in front of a small object, and look at that object, I see two fingers; or if I look at my finger, I see two objects. Explain this.

9. Enumerate the constituents of white light. How was this discovered, and by whom?

10. What is meant by chromatic aberration? How may it be obviated?

11. Sketch a prism. Show the action of a prism upon a ray of homogeneous light.

12. Point out some analogies in the phenomena of sound, light, and radiant heat.

HEAT.

CHAPTER I.

76. Nature of Heat.—As in the case of light, there have been two theories brought forward to explain the phenomena of heat. One is called the *material theory*, and the other the *dynamical theory*.

According to the material theory, heat is a *kind* of matter; that it consists of an imponderable substance surrounding the molecules of bodies, and that in virtue of its attraction for other matter, and its repulsion for its own particles, it can readily pass from one body to another. According to the dynamical theory, heat is an *affection* or *condition* of matter; that the heat of a body is caused by a vibratory motion among its particles; further, that the molecules of warm bodies possess the power of communicating a vibratory motion to the surrounding *ether*, in virtue of which contiguous bodies may be heated. Very hot bodies are, on this hypothesis, those which give a rapid vibratory motion to the all-pervading ether. The latter theory is the one more generally accepted by modern physicists.

77. Heat and Cold.—The ordinary sensations which are familiarly known as *heat* and *cold*, are merely different degrees of the same influence. If we touch in succession, for example, two similar bodies unequally heated, the one may appear to the hand cold, the other hot; but these sensations are due to the fact that the bodies have different amounts of the *same* influence, viz., *heat*.

78. General Effect of Heat.—The most general effect

of heat on a body is to expand it, or cause an enlarge-
ment of its volume. If heat be abstracted from a body,
or the body be cooled down, the contrary effect ensues—
the volume of the body is diminished. Hence we have
the general principle (to which, however, there are some
exceptions), *that heat expands and cold contracts.* These
effects, arising either from an increase or diminution of
temperature, are produced in very different degrees, according to the nature of the bodies. They are small in
solids, greater in liquids, and still greater in gases.

79. Expansion of Solids.—The expansion which a
body undergoes, in regard to *length*, is called *linear ;*
in regard to *surface, superficial ;* and in regard to
volume, cubical, expansion.

To illustrate linear expansion, the following apparatus
has been devised (fig. 67):—

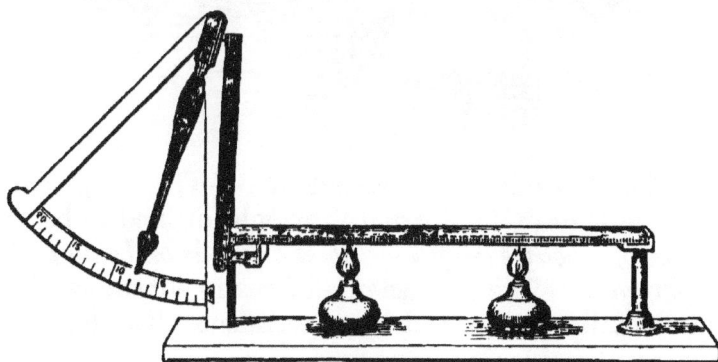

Fig. 67.

A metallic bar is supported on pillars, as in the
figure. Its free end presses upon a lever, which in
turn acts upon an index, playing over a graduated
arc. By this arrangement a multiplying effect is produced upon the index, in consequence of which the
smallest elongation on the part of the bar is rendered
manifest.

The cubical expansion of a body is shown very simply
thus : A small brass ball (fig. 68) just passes through
an aperture in a metal plate supported on three legs.

When the ball is heated by being held over a spirit lamp, it refuses to go through the aperture, and will not do so until it regains its former temperature.

Fig. 68.

80. The Co-efficient of Expansion.—The *co-efficient* of expansion (linear, for instance), may be defined to be that *fraction of a body's length which it expands on being heated* 1° *centigrade.* The co-efficients of many substances have been carefully determined by experiment. The following table may be given as a specimen :—

CO-EFFICIENTS OF EXPANSION (LINEAR).

Zinc	·0000294	Gold	·0000146
Silver	·0000190	Iron	·0000123
Brass	·0000188	Platinum	·0000088
	Glass	·0000080	

It is easily proved that the superficial is *double* of the linear co-efficient, and that the cubical is *treble*.

From this table it appears that zinc is the most expansible metal, and platinum the least; also that the expansibility of platinum and glass is nearly the same. Hence

8 E   F

the reason why a chemist can fuse a platinum wire into a glass tube without liability to fracture.

81. Practical Applications.—The principle of expansion or contraction is utilized much in practice. The hoop of iron by which a wheel is surrounded, is made of the same diameter as the wheel. It is then heated, and in this state is put on the wheel. The whole being thrown into water, the iron hoop contracts with great force, and thus binds the spokes and rim firmly together. A similar method is employed for binding together the staves of tubs, vats, barrels, etc. The walls of a building have been restored to their perpendicular position by taking advantage of the enormous contractile force of iron.

In the combination of metallic pipes, by which water is brought from great distances for the supply of towns, means must be provided for allowing expansion or contraction to take place freely. Hence the pipes are so constructed as to be capable of sliding one within the other, after the manner of the joints of a telescope. In all iron bridges, similar precautions are necessary; they are generally supported on friction rollers.

The same principle explains certain familiar facts. Thus, when hot water is poured into a cold glass vessel, fracture often takes place. This arises from the *unequal* expansion of the glass, the heat not having had sufficient time to extend its influence equally to other parts of the vessel. The same accident may take place when cold water is poured into a warm glass vessel. When the stopper of a decanter becomes firmly fixed, it is not unusual to wrap a cloth steeped in hot water round the neck, the neck thereby expands, and the stopper is freed from its hold.

82. Breguet's Metallic Thermometer.—That metals exhibit different amounts of expansibility may be illustrated by the following experiment :—Two strips of iron and brass are firmly riveted together (fig. 69). At the ordinary temperature the combined strip is perfectly straight, but when heated it becomes bent. If cooled below its ordinary temperature it also becomes bent, but in the opposite direction. The experiment demonstrates, therefore, the greater expansibility or contractibility of brass than iron.

On the same principle is founded Breguet's metallic thermometer, represented in fig. 70. Three strips of silver, gold, and platinum, are rolled into a very thin metallic ribbon. This ribbon is coiled into a spiral form, and adjusted as in the figure, the internal face being the silver and the external the platinum. As the temperature rises the spiral unwinds itself; as it falls, it moves in the opposite direction, and these changes affect the index which plays over the graduated circle.

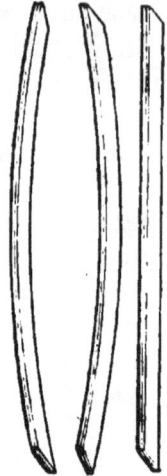

Fig. 69.

83. Gridiron Pendulum.—In the finer kinds of clocks the variation of temperature is guarded against by the use of what is called a *compensation* pendulum. A common form is the "gridiron" pendulum represented in fig. 71. It consists of a combination of steel and brass rods, S, B, arranged alternately, and of such length as that the expansion or contraction of the steel rods may be exactly neutralized by the expansion or contraction of the brass ones. To the middle steel rod is attached the bob C.

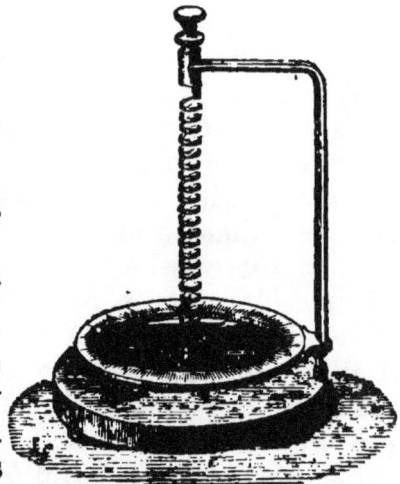

Fig 70.

To illustrate its action, let O be the centre of oscillation of the pendulum, or, which is the same thing, let AO be the length of the equivalent *simple* pendulum. In

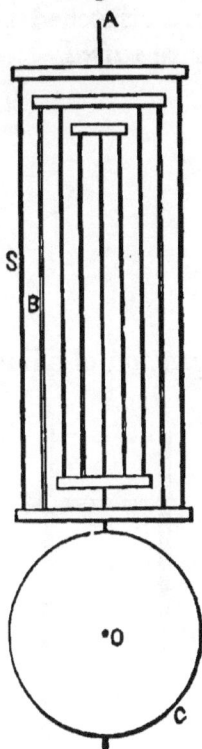

Fig. 71.

summer the steel rods will expand, and thus tend to lower the point O, or lengthen the pendulum; but the brass rods also expand, and, by their so doing, they tend to raise the point O, or shorten the pendulum. If, therefore, the point O is as much raised by the expansion of the brass rods as it is depressed by the expansion of the steel ones, it will be kept in the same position, in other words, the length AO will be unchanged. In winter, in like manner, if the effects of contraction in the one case be equal to the effects of contraction in the other, the length of the pendulum will be preserved. Hence, by this arrangement, the pendulum is kept constant in its length.

84. Exceptions to Expansion.— There are some *exceptions* to the principle of expansion by heat which are worthy of notice.

In what is called "Rose's fusible metal," which is an alloy of bismuth, lead, and tin, in certain proportions, the expansion is nearly uniform from 32° to 110° Fahrenheit, but at this point the expansion ceases, and as the temperature rises to about 155° it undergoes a constant contraction.

There are some crystals which, when heated, expand in one direction, but *contract* in another.

But, perhaps, the most singular exception is found in the case of india-rubber. A piece of stretched india-rubber, on being heated, contracts. This substance also forms an exception to the all but general rule, that when a body is stretched, *cold* is developed.

If a wire, for example, be lengthened, its temperature is lowered; not so with india-rubber, a stretched piece of rubber is found to have its temperature *raised*.

CHAPTER II.

85. Expansion of Liquids.—The expansion of a liquid, such as water, may be proved by the following experiment: Take a narrow tube (fig. 72.) with a large bulb at its extremity, fill the bulb and part of the tube with coloured water. If now the bulb be heated by a lamp the water begins to expand, and, owing to the great difference of capacity between the bulb and the tube, any small expansion is rendered manifest by a considerable rise of the liquid in the tube; and in a short time it will reach the top.

The *co-efficient* of expansion of a liquid (having reference of course to *cubical* dilatation only) is, as before, that fraction of its volume which it expands on being heated 1° C.

Fig. 72.

Of all liquids which have been subjected to experimental investigation, there are none which have been more thoroughly examined than water, alcohol, and mercury.

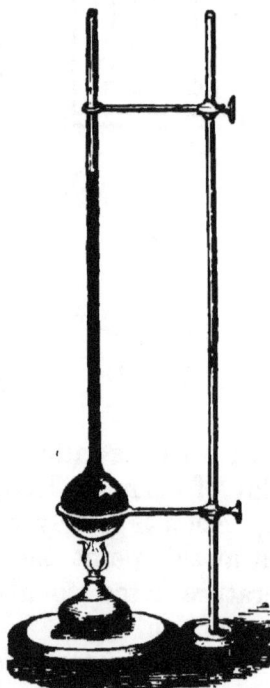

The following co-efficients have been determined in regard to them :—

CO-EFFICIENTS OF EXPANSION BETWEEN 0° AND 100° C.

Alcohol	·00116
Water	·000466
Mercury	·000154.

The expansion of these liquids, however, is not perfectly uniform; there is found to be great irregularity near their boiling points. In the case of mercury, its expansion is nearly uniform between the limits of – 36° C. and 100° C.

86. The Thermometer.—In consequence of the uniform expansion of mercury, and its great sensitiveness to heat, it is the fluid more generally used in the construction of the *thermometer*—the common instrument for indicating temperature.

The mercurial thermometer* consists of a tube with a small uniform bore, terminating in a bulb. The bulb and part of the tube are filled with mercury, in such quantity as that the mercury may neither pass wholly into the bulb, nor reach the top of the tube, when subjected to the ordinary extremes of cold or heat. The graduation of the instrument is effected in the following manner :— Two invariable or fixed points are selected, viz., the *freezing point*, and the *boiling point* of water at the sea level, and under the mean pressure of the atmosphere. The former point is obtained by plunging the tube into melting ice, when the column gradually sinks, and eventually comes to rest; the lowest level of the mercury is then marked off on the scale attached to the tube. The latter point is determined by placing the instrument in boiling water, or in the steam escaping from it; this being done the mercury rises, and finally settles at a certain height — the highest level attained is marked off as before. The interval between the two fixed

* For the construction of the instrument, see Appendix, Question 12, p. 139.

points thus determined is then divided into so many equal parts.

87. Thermometric Scales. — The interval just mentioned is differently divided in different countries, giving rise to the three common forms of the thermometer. These are named from the inventors, and are known as the *Fahrenheit, Celsius* or *centigrade,* and *Réaumur.* In Fahrenheit (fig. 73) the freezing point is marked 32 (a zero being taken which was incorrectly imagined to be the greatest cold obtainable), and the boiling point 212; in centigrade these points are marked respectively 0 and 100; and in Réaumur, 0 and 80. The space therefore between the two fixed points is divided in F. into 180 equal parts, in C. into 100, and in R. into 80. Each of these parts is called a degree.

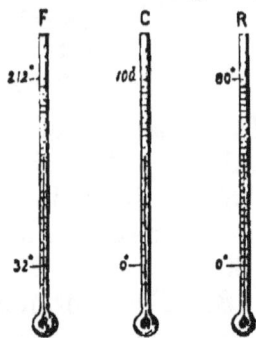

Fig. 73.

A temperature below 0° in any of the scales, is indicated by a minus placed before the number. Thus – 10° C., indicates 10 degrees below the freezing point according to the centigrade scale; and again, – 10° F., 10 degrees below 0° according to the Fahrenheit scale.

88. Conversion from one Scale to another—Examples. —The student will have little difficulty now in understanding the mode of converting any number of degrees of one scale into its equivalent in another. The following examples may be given in illustration :—

Ex. 1.—*Convert* 68° F. *into the centigrade scale.*

The freezing point being marked 32 in Fahrenheit and 0 in centigrade, we must make allowance for this by first subtracting 32 from 68; the remainder is 36. Now as 180° F. = 100° C., we have therefore to state (by the common rule of three) as follows:—180 : 36 :: 100 = 20; hence 68° F. = 20° C.—Ans.

Ex. 2.—*How many degrees of F. are equal to* – 40° C. ?

Here we must state the proportion thus:—
100 : − 40 :: 180 = − 72. *Adding 32 we have*
− 72 + 32 = − 40; *hence* − 40°C. = − 40° F.—Ans.

Ex. 3.—*How many degrees of R. are equal to* 100° F.?
*100 − 32 = 68. *Then* 180 : 68 :: 80 = 30$\frac{2}{9}$°;
hence 100° F. = 30 $\frac{2}{9}$°R.—Ans.

Ex. 4.—*Convert* − 30° R. *into Fahrenheit.*
80 : − 30 :: 180 = − 67$\frac{1}{2}$. *Adding* 32* *we have*
− 67$\frac{1}{2}$ + 32 = − 35$\frac{1}{2}$; *hence* − 30° R. = − 35$\frac{1}{2}$° F.—
Ans.

89. Ebullition.—The process by which water is raised

Fig. 74.

to the boiling point is a very interesting one. Thus, let an open flask of water be exposed to heat, as in fig. 74. The stratum of fluid at the bottom, in becoming heated, expands and rises towards the surface; another stratum taking its place in like manner expands and rises, and so on successively. There are produced therefore in the vessel a series of ascending warm currents and of descending colder currents, and this circulation continues until the water is nearly brought to the boiling point. When this point is reached, bubbles of gas are observed to form themselves next the heating source. These at first, in their passage upwards through the colder

* The student will notice therefore that in the conversion of Fahrenheit degrees into centigrade or Réaumur, he must first *subtract* 32 from the number given *before* stating the proportion, and in the converse problem *add* 32 *after* he has worked the proportion.

water above, are gradually condensed, and diminishing in volume as they ascend scarcely reach the surface; but in proportion as the *whole* water approaches the boiling point this condensation ceases, and the bubbles escape at the surface as steam. The water is then said to *boil*.

90. The Dependence of the Boiling Point upon External Pressure.—The temperature at which water boils in an open vessel is dependent upon the pressure of the atmosphere. At the ordinary pressure, that is, when the barometer indicates about 30 inches of mercury, the boiling point is 212°F. But if the pressure diminish the boiling point falls, and, on the other hand, if the pressure increase it rises above the temperature of 212°. Hence the necessity of strictly defining what the *boiling point* of a liquid really is. It is *that point of temperature at which the tension or elastic force of its vapour is exactly equal to the pressure it supports.*

The variation of the boiling point of water with the pressure will be seen from the following table :—

Height of the Barometer. (Inches).	Boiling Point. (Fahrenheit).
17·04	185°.
18·99	190°.
21·12	195°.
23·45	200°.
25·99	205°.
28·74	210°.
29·33	211°.
29·92	212°.
30·51	213°.
31·73	215°.

From this table it appears that a variation of $\frac{1}{10}$ of an inch of the barometer causes a difference of about $\frac{1}{6}$ of a degree Fahernheit in the boiling point; hence the range of the boiling point in our climate may be as much as 5°, with the ordinary variations of the barometer.

In a *closed* vessel, water may be raised to a much higher temperature than 212°. This is the case in the

boiler of a steam-engine, or of a locomotive. By the
accumulation of the steam the pressure on the water is
increased, the boiling point is therefore raised, or the
water is heated above its ordinary boiling point.

91. Illustrations.—A striking illustration of the dependence of the boiling point upon external pressure, is
to take a vessel of hot water, put it under the receiver of
the air pump, and exhaust the air. In a short time the
water begins to boil, and as the rarefaction goes on, the
ebullition increases in intensity.

Another experiment consists in taking a vessel of hot
water (fig. 75), corking it up, and then inverting it. If now

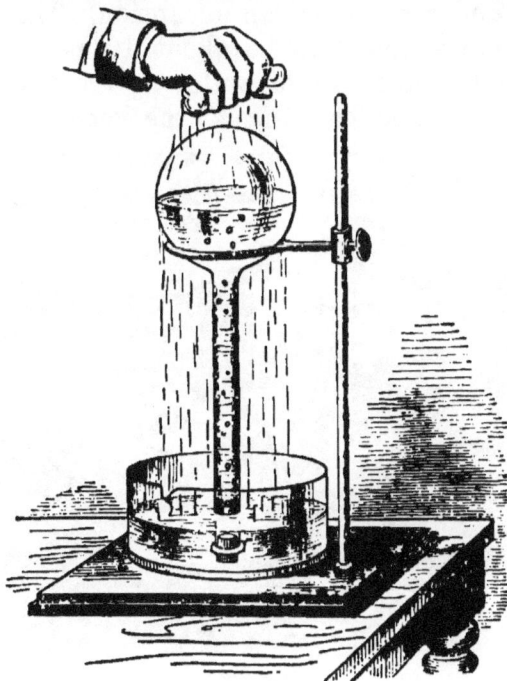

Fig. 75.

cold water be allowed to fall over the confined vapour, it
partially condenses it; the water in the vessel therefore is
so far relieved from pressure, and in consequence enters
into a state of ebullition.

92. Maximum Density of Water.—If a quantity of water, say at the temperature of 62°F. (standard temp.), be gradually cooled down, it contracts until it reaches the temperature of 39°, when all further contraction ceases. This point is called the point of *maximum density.* When cooled below this temperature *expansion* sets in, which increases rapidly as the freezing point is approached. Water is therefore *heaviest* at the temperature of 39° F. or 4° C. For example, a cubic foot of water at that temperature weighs more than a cubic foot at any other temperature.

93. Deportment of Water in Freezing.—When water freezes, it undergoes a sudden expansion. The amount of its expansion is found to be about 10 per cent.; more exactly, 1000 cubic feet of water at the freezing point become 1102 cubic feet of ice at the same temperature.

Fig. 76.

The force of this expansion is almost irresistible. A strong iron bottle filled with water, and firmly closed, when immersed in a freezing mixture, is rent asunder in a short time. Some interesting experiments on this point were made one severe winter at Quebec by Major

Williams. He took a bomb-shell, and having filled it
with water, carefully plugged up the aperture; on exposing it to the frost, the plug was driven to a distance of
330 feet, whilst at the same time a cylinder of ice $8\frac{1}{2}$
inches long appeared protruding at the aperture (fig. 76).
In another experiment, the plug being more firmly fixed,
the bomb was ruptured at the middle, and a ring of ice
was forced through the rent.

The common accident of the bursting of pipes in frosty
weather, can therefore be easily understood. The rupture
takes place, of course, during the frost; but the rent being
closed up with ice, no leakage of water takes place. It
is only when the thaw sets in that the damage done to
the pipe becomes apparent.

94. Effects in Nature.—In the economy of nature the
expansion which accompanies the freezing of water exerts
a most important agency. Had water, in cooling, observed
the general law of contraction, then a layer of ice when
formed on the surface of our lakes or rivers would have
sunk to the bottom; another would have been formed,
and in like manner have sunk to the bottom, and so on
until the whole water had become one solid mass of ice,
which all the influence of a summer sun could scarcely
have dissolved. As it is, however, these effects are
happily prevented. The ice being lighter than the water
floats on the surface, and thus the water below, being
sheltered from the cold atmosphere above, preserves its
liquid form.

It is thus also that our soils are pulverized during
winter. The water they imbibe, upon freezing, disinte-
grates them, and thereby assists, in no small degree, the
labours of the husbandman in preparing them for the re-
ception of the seed. Hence, during frost, the soil is
observed to have a cracked appearance.

**95. Expansion on Solidification—a Property not
Peculiar to Water.**—There are certain metals which,
in passing from a state of fusion into the solid form,
manifest the same deportment as water; they expand on

solidifying. These are, cast-iron, bismuth, and antimony. Hence the precision with which cast-iron takes the impression of a mould.

CHAPTER III.

96. Expansion of Gases.—The expansion of gases exceeds that of either solids or liquids, and is *almost* uniform; that is, the amount of expansion is found to be nearly in proportion to the increase of temperature. The process of dilatation, or contraction of gases, does not take place in the same manner as in solids and liquids. Let us suppose, for instance, that we have a glass receiver closed on all sides, and filled with air of the same density as that of the external atmosphere. If the temperature of this enclosed air be lowered, it will not contract in its dimensions, it will still occupy the *whole* of the receiver, but its elastic force is reduced; in other words, it will not exert the same pressure on the containing vessel as before, and were the external pressure of the air allowed to act, it would force the confined air into a smaller space. In like manner, suppose the air contained in the receiver to be subjected to an increase of temperature, then the elastic force of the air is increased, and were the receiver to offer no resistance, it would expand and occupy a greater space. By gases expanding or contracting by a change of temperature, is therefore meant that they do so under a *given pressure*—the pressure generally taken being the *ordinary pressure of the atmosphere.*

97. Experimental Illustrations. — (1) A bladder partly filled with air and closed up, when held before a fire, becomes gradually inflated, shrinking to its former dimension on its removal. (2) When a flask of water, as in (fig. 73), is at first heated, bubbles of air are seen to rise through the water, owing to the expansion of the air-particles which have been absorbed by the water. This is more strikingly seen with ale or other fermented liquor: a quantity of froth collects on the surface in proportion as the gaseous particles are liberated. Hence, when a bottle of such liquid is placed before a fire, it often happens either that the bottle is broken, or the cork driven out with a loud report. (3) A flask, A, containing air, is taken (fig. 77), from which a bent tube is led to a dish

Fig. 77.

B filled with coloured water. Over the end of this tube is placed an upright tube C, previously filled with the fluid and inverted. If now the flask be heated by a spirit lamp, the air inside expands, passes through the bent tube, and collecting in C, gradually displaces the fluid, and eventually expels it entirely.

98. Fire-Balloon.—In the experiment just mentioned, it is manifest that the air thus expanded is lighter, bulk for bulk, than the air of the external atmosphere. If therefore we were to take a balloon and fill it with heated air, it would ascend and remain in an elevated position so long as the heat of the air is preserved. Such is the principle of the so-called "fire-balloon."

99. Constancy of the Co-efficient of Expansion.—The co-efficient of expansion is found to be *nearly the same in all gases.* Its amount may be taken at ·00366, or about $\frac{1}{273}$ of its volume. Thus 1 cubic foot of gas at 0°C. becomes $1\frac{1}{273}$ cubic feet when raised 1°C., $1\frac{2}{273}$ cubic feet 2°C., $1\frac{5}{273}$ cubic feet 5°C., etc. (the pressure on the envelope containing the gas being preserved constant). Slight deviations from the general rule of constancy in the co-efficient occur in the cases of *carbonic acid* and *sulphurous acid* gases. This is owing probably to their capability of being liquefied.

100. Physical Character of Carbonic· Acid and Sulphurous Acid Gases—

(1) *Carbonic Acid* (CO_2). This gas is one-half heavier than common air, it is colourless, but possesses a slight odour, and a perceptibly sour taste. Its principal chemical feature is that it extinguishes flame, and causes death to an animal inhaling it. It is present in the atmosphere, and in the water of many mineral springs. The quantity present in free air is nearly constant, amounting to about 4 volumes in 10,000 of air. Small as this appears to be, it is nevertheless sufficient and necessary for the support of vegetable life.

Carbonic acid is given off by animals in respiration and by combustion. Fermented liquors, soda water, etc., owe their sparkling briskness to the escape of this gas. It may be liquefied at a pressure of about 36 atmospheres. The liquid possesses the remarkable property of being more expansible than the gas itself—a strange exception to the rule that liquids expand less by heat than gases.

(2) *Sulphurous Acid* (SO_2) is given off when sulphur is burnt, and in large quantities from volcanoes. It is colourless, but possesses a suffocating smell of burning sulphur. It is $2\frac{1}{4}$ times as heavy as air, and is reduced to a liquid at a pressure of two atmospheres, or by being

cooled down to $-10°C$. under the ordinary atmospheric pressure.

101. Draft of Chimneys—Ventilation.—When a fire is kindled in a room, the flame and warm smoke proceeding from it soon raise the temperature of the air in the chimney. The consequence is it ascends, and the colder air from the room flows in to supply its place; this air, in turn likewise becoming heated, rises, and a fresh accession of air takes place, and so on.

What constitutes, therefore, the *draft* of a chimney is nothing else than the colder air of the room constantly passing towards the fire-place.

As the air in a room is continually passing towards the fire, there must of course be a constant supply kept up from the external air, which must therefore have sufficiently free access by the doors and windows of the house. Hence it is found that in a house where the passage of the external air is much interrupted, the chimneys are liable to smoke, the reason being that a sufficient draft is not maintained.

At the door of a room where there is a fire there are two opposite currents of air, the heated air in the room ascends to the top and passes out at the upper part of the door, whilst the colder air from without enters by the lower part. This may be easily proved by placing a lighted taper in these positions at the outside of the room door. In the former position the flame is bent *from* the door, and in the latter *towards* it.

When all the windows and doors of a house fit so closely as to impede a communication with the external air, and thus prevent a sufficient supply for the fires in the house, the necessary quantity descends by those chimneys which are not in use. Hence, when a fire is being lighted in any of these, the smoke at first is driven into the room. To remedy this the room door ought to be shut, or the window opened; this being done, the chimney will soon begin to *draw*. What is called *back smoke* in a room where there is no fire arises from the circumstance that that chimney is serving as an inlet for air to supply the fires in the house, carrying the smoke of a neighbouring chimney down into the room along with it.

The grand object in *ventilation* is to allow the heated air, or air vitiated by respiration, to escape at the roof of the building, whilst provision is made at the same time for the inlet of fresh air, the whole arrangements being such as to obviate drafts. The principle of ventilation is strikingly illustrated by the following simple experiment: A glass receiver (fig. 78) with an aperture at the top, is placed over a candle put into a flat dish, in which there is water. In a short time the air in the receiver becomes vitiated by the combustion, and the candle flame, gradually dwindling down, is at last extin- guished. If, however, the candle be relit and a card be placed in the funnel, or chimney, thus dividing it into two parts, the candle continues to burn, preserving its brightness almost unimpaired. The

Fig. 78.

reason of this is, that the vitiated air now escapes through one of the passages, whilst fresh air gets in by the other, as indicated by the arrows.

102. **Winds—General Character of.**—The phenomena of *winds*, in general, result from *the unequal distribution of heat over the earth's surface.* It is ascertained, for example, that the *mean* temperature at the equator is 84°F., at 78° north latitude 16°F., and at the pole it has been inferred to be about 4°F. From such diver- sity, then, of temperature in different portions of the earth's surface, it is impossible that the atmosphere can

8 E   G

remain calm and unaffected. When any region becomes
more heated by the sun's beams than some other, the
air above it also gets heated and rises up, whilst the
colder surrounding air rushes in to supply its place, and
the atmosphere is more or less disturbed.   Hence arise
those perturbations to which we give the name of
" winds."

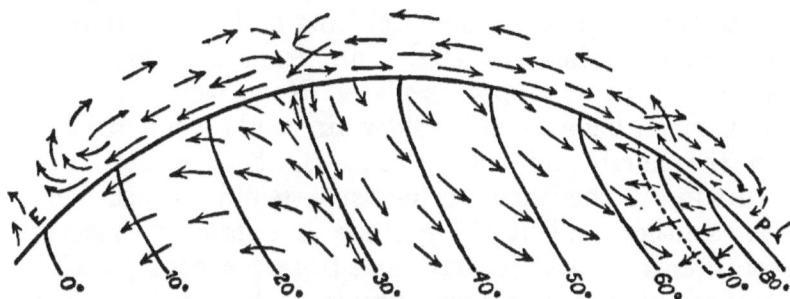

Fig. 79.

The accompanying illustration (fig. 79) gives a general
view of the character of the winds which prevail in the
northern hemisphere, from the equator to the pole.   The
arrows show the direction of the aërial currents.   The
warm air from the tropics, ascending to a certain height in
the atmosphere, flows northward as an upper current; on
cooling down it descends about the 30th parallel of latitude, and blows as a south-west wind between that
parallel and the 60th ; getting warm by contact with the
earth's surface, it again ascends, still flowing towards the
pole, where it at length precipitates itself and forms the
polar gales. Returning now southwards, it ascends at the
60th parallel, blowing as an upper current, till, on getting
chilled, it descends at the 30th parallel, and, between that
and the equator, blows as a north-east wind.   Thus a
continuous circulation of air goes on.

It must be understood, however, that these aërial
currents are subject to considerable variation, swayed as
they are by a number of disturbing influences, which.
more or less affect them.

103. Trade Winds—Land and Sea Breezes.—The

trade winds, so named from their importance to navigation, and hence to the purposes of trade, are those which prevail in the tropics. They blow in the northern hemisphere from the *north-east,* and in the southern hemisphere from the *south-east.* They are easily accounted for. In consequence of the high temperature of the tropics, there is a continual uplifting of heated air from that region, whilst the colder air from the temperate zones rushes in to supply its place. Now, were the earth stationary on her axis, this colder air would come directly from the north and south; that is, there would be a north wind in the northern hemisphere and a south wind in the southern. But the earth is in a state of constant revolution from west to east, and from her configuration it is clear that the different points in her surface have very different rates of motion. The colder air, therefore, in its passage towards the equator, will move over latitudes which are gradually increasing in velocity. It cannot acquire all at once the velocity of that part of the earth over which it is advancing. It must necessarily lag behind, and be *struck* by the objects in that zone with a certain force. Thus it is that air is influenced by two motions; *first,* a northerly or southerly motion, caused by its tendency to rush to the equator to supply the place of the heated air there; and *second,* an easterly motion, caused by the rotation of the earth.

By a well-known principle in dynamics * the air will obey neither the one motion nor the other, but will take an intermediate course. In other words, the wind will blow from the north-east in the one case, and south-east in the other.

The following familiar illustration may be given:

* The principle here referred to is the "parallelogram of motion." If a point A is urged by a force which would make it move over the space A B in a certain time, and by another force over A C in the same time, then completing the parallelogram, the point under the influence of the two forces will move through the diagonal A D.

Suppose a horseman to gallop along towards the east on a quiet day; his motion will give rise to a certain resistance on the part of the air, and he will imagine that an east wind is blowing; but let him do the same thing when a slight north or south wind is blowing, his motion will give an easterly character to either wind, and he will judge of the wind as coming from the north-east or south-east. There is a similar effect in the trade winds.

The *land* and *sea breezes* are peculiar to maritime localities. During the night the land loses its heat by radiation more rapidly than the sea, the cool air from the land therefore makes its way to the sea, displacing the warmer air there. This constitutes the land breeze. During the day the reverse takes place; the cool air from the sea flows to the land, forming the sea breeze. The beneficial effects of the sea breeze are particularly felt in tropical climates. Many islands would be almost uninhabitable were it not for the sanitary influence of this breeze.

CHAPTER IV.

104. Aqueous Vapour — Evaporation.—The atmosphere, however clear and pure it may appear, always contains a quantity of watery vapour, existing as *an impalpable transparent gas*. The amount varies in different portions of the atmosphere, but is generally small. Thus, in 100 parts of ordinary air there are found to be only ·45 parts of aqueous vapour, the remaining 99·55 parts consisting of the other constituents, oxygen, nitrogen, and carbonic acid gases. How does this aqueous vapour originate? By the slow and mysterious process of *evaporation*.

That this process is in constant action is proved from various familiar facts. A tumbler of water put outside a window in dry weather gradually loses its contents. No sooner are our streets watered than they begin to dry. Wet clothes, in like manner, hung up in an open place soon lose their moisture. The rate at which evaporation takes place depends upon the temperature of the air. Even at a low temperature we find evaporation still going on. A piece of ice, for example, when exposed to severe cold, gradually diminishes in size, and eventually disappears. On the other hand, as the temperature is increased, evaporation takes place more readily. Hence the rapidity with which our streets are dried in summer after a shower of rain. Hence also the diminished size of our rivers in summer after a continuance of fine weather.

105. Point of Saturation.—It can readily be imagined, therefore, that even throughout the same day evaporation will take place at different rates, resulting from the varying heating power of the sun, and therefore that the quantity of aqueous vapour carried up will also vary. Thus, though there be considerable difference in the capability of the air as regards the suspension of aqueous vapour for different temperatures, there is for each temperature a definite quantity which can be elevated. When air has reached this state it is said to be *saturated*, or to have attained its *point of saturation*. Thus, if there are two equal masses of air, of 40°F. and 70°F. respectively, each can take up its own definite quantity of vapour; but the former, owing to its lower temperature, will reach its point of saturation sooner than the latter.

106. Air Heated by Compression and Chilled by Expansion.—When air is compressed heat is evolved. This can be shown by taking a brass cylinder with a piston fitting it air-tight (fig. 80). In a small aperture at the end of the piston-rod is inserted a piece of tinder. If now the piston be forced down into the cylinder, the air inside becomes compressed, and sufficient heat is evolved to kindle the tinder.

An instrument of this kind has been long in use among some of the native tribes of India.

. Again, when air is rarefied or expanded cold is produced. A striking proof of this is afforded when a
receiver is being exhausted by
an air pump. After one or two
strokes a *cloudy* appearance is
observed in the receiver, resulting from the condensation of the
suspended vapour in consequence
of the air being chilled.

107. Clouds — Rain. — If a
heated mass of air, charged with
aqueous vapour, be carried aloft,
it will expand by reason of the
diminished pressure upon it, and
become chilled. It can no
longer, therefore, hold all its
vapour in suspension ; condensation sets in, and that the more
rapidly as it ascends, hence the
formation of a cloud. *Clouds,*
therefore, are masses of aqueous
vapour in a partially condensed
state. They are not so high as
they appear. The greater number
of clouds we see are within a
few thousand feet of the earth's
surface. Hence a mountain
traveller often becomes enveloped
in clouds, or if he has attained
a considerable elevation, he may

Fig. 80.   witness some clouds floating below him. The motion of clouds is not so regular as we
are apt to suppose ; they have not a motion of trans-
ference merely, but also one in a *vertical* direction,
arising from the continued and variable effects of ascending currents.

If the condensation of the vapour in the atmosphere
be not confined to the higher regions, but is spread over

the surface of the earth, then there is a *mist* or *fog* formed. Fogs arise, for the most part, from the surfaces of rivers, or lakes, or from the damp ground being *warmer* than the superincumbent air.

Rain is caused by a considerable diminution in temperature, and therefore by a rapid condensation of the aqueous vapour. A cloud is capable of holding its moisture so long as the temperature keeps sufficiently high, but should it be carried by the wind into a cool region, it becomes no longer able to do so; rapid condensation sets in, vesicle unites with vesicle, and rain falls. At first the drops are small, but they gradually increase in size, from their uniting with other vesicles in their descent. The rapidity of the rain-fall depends upon the amount of vapour in the cloud, and upon the decrease of temperature to which it is subjected. If these elements are carried out far, then the rain-fall will be correspondingly great, hence the heavy rains in thunderstorms.

108. Dew—Dew-point—Hoar-frost.—The phenomenon of *dew* affords a vivid demonstration of the constant presence of aqueous vapour in the atmosphere. It is easy of explanation. After sunset, the different objects on the earth's surface begin to radiate or part with the heat which they have absorbed during the day; as the night advances this radiation proceeds, until at length they acquire a much lower temperature than the air above them. The consequence is that the air gets chilled *below* the point at which it can hold its vapour in suspension, condensation ensues, and a deposition of moisture or dew takes place. The point at which the deposition begins is termed the *dew-point*. The amount of this deposition on the different objects depends upon their radiating powers. Thus, dew is found to form copiously on grass, the leaves of flowers or trees, and on other products of vegetation, because these are good radiators of heat; whilst, again, the supply is small on stones or the naked soil, because their radiating power is feeble.

It is only, however, in certain states of the atmosphere that dew is deposited. Cloudy or windy nights are unfavourable to its production. In the former case, though there is radiation going on from the earth's surface, yet the clouds are also good radiators, and they thus prevent the surface from being cooled much below the temperature of the atmosphere; in the latter case, the constant transfer of the air from place to place acts as a preventive. Clear still nights, on the other hand, are the most favourable, for then the radiation goes on freely.

Hoar-frost is just dew in a frozen state. The formation of hoar-frost is therefore entirely influenced by the causes affecting the deposition of dew.

109. Snow—Snow-Crystals.—When the temperature of the air is below 32°F., the vesicles of vapour become frozen, and in uniting together become heavier than the air, and fall as *snow*. The flakes are sometimes small, at other times large, their size depending upon the amount

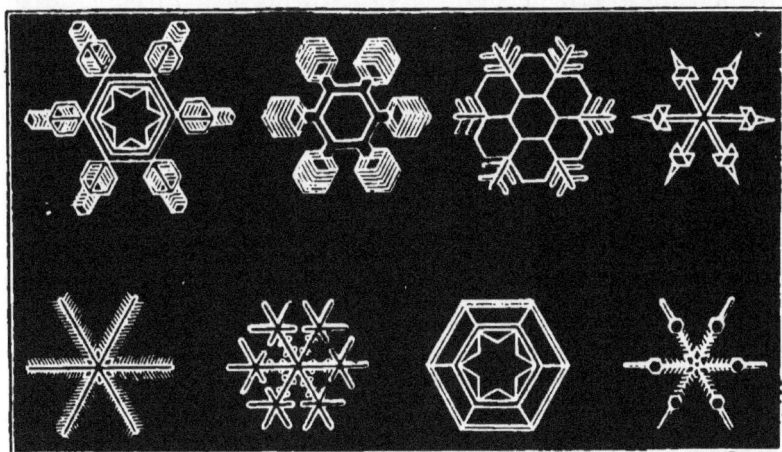

Fig. 81.

of moisture and the extent to which the low temperature prevails. Should the flakes, on their descent, encounter warm strata of air, a partial fusion takes place, and they fall in a half-melted state, forming *sleet*.

Examined with the microscope, snow presents a very

beautiful appearance; it is formed of a number of distinct and transparent crystals of ice, which are observed to be grouped together in a variety of ways. Fig. 81 exhibits some of the different forms of snow-crystals which are found.

110; Hail.—*Hail* may be regarded as frozen drops of rain. A small hard nucleus, or centre, is first formed in the upper regions of the atmosphere; this, on its descent, collects more and more moisture on the surface and freezes it, till it at length falls, of some magnitude. Hail rarely falls in winter, chiefly in spring and summer. In winter, from the prevalence of a low temperature, the vapour is condensed and frozen *before* the particles can unite to form drops; hence, in that season, we have snow but not hail. In spring and summer we have often electric discharges, and as these sometimes produce a very sudden cold in the region of the atmosphere where they occur, such discharges are not unusually accompanied by a fall of hail. Hail-storms are often a great scourge to the agriculturist.

CHAPTER V.

111. Specific Heat.—*The specific heat* or *capacity for heat* of a body, is the quantity of heat necessary to raise it through a certain number of degrees, as compared with the quantity required to raise an equal weight of water through the same number of degrees. In this country it is customary to express by *unity* the amount of heat necessary to raise one pound of water 1°C., or, which is the same thing, the amount of heat which one pound of water gives out in falling 1°C. This is known as the *thermal unit*. For example, take water and mercury. It is found that thirty times as much heat is required to raise one pound of water 1°C. as is required to raise one pound of mercury 1°C. If, therefore, we express the specific heat of water by 1, we must express the specific heat of mercury by $\frac{1}{30}$ or ·03.

112. Methods of Measuring the Specific Heat of Bodies.—We mention two methods of measuring specific heat.

(1) *Method of Mixtures.*—This consists in placing a given quantity of the substance (whose specific heat is required) at a given temperature, in a given quantity of water at a lower temperature, and ascertaining the loss of heat by the former, and the gain by the latter. An example will illustrate the method : Suppose we mix 5 lbs. of a fluid (call it A) at 80°C., with 2 lbs. of water at 10°C., and that the temperature of the mixture is 25°C.; denote by x the specific heat of the fluid. We have here a *decrease* of temperature in A of 55°, and an *increase* in the water of 15°. Therefore the amount of heat given out by 5 lbs. of A will be expressed by $5 \times 55 \times x$; whilst the amount of heat absorbed by the 2 lbs. of water will be expressed by $2 \times 15 \times 1$. Then, since the loss of heat in the one case is just equal to the gain in the other, we have $5 \times 55 \times x = 2 \times 15$, and $x = \frac{2 \times 15}{5 \times 55} = ·109$ nearly, hence the specific heat of A = ·109.

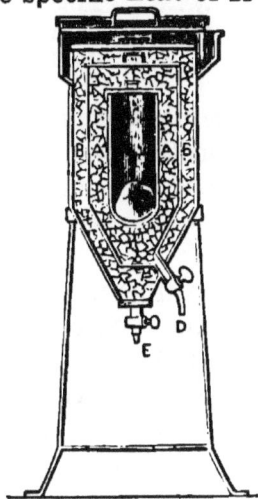

(2) *The Ice Calorimeter.*—This instrument was invented by the French philosophers, Lavoisier and Laplace. A sectional drawing of it is shown in fig. 82. It consists of three tin vessels, one within the other, the spaces A, B, between being filled up with pounded ice at 0·C°. The body, whose specific heat is to be determined, is placed in the central one. There are two stop cocks E, D, for running off the water caused by the fusion of the ice on the part of the surrounding atmosphere and the heated body

respectively. In order to use it, the body of weight W, suppose, being raised to a given temperature t, is quickly placed in the central vessel, and allowed to remain there till its temperature sinks to 0°C. The water resulting from the fusion of the ice is then drawn off at the stop cock D and weighed. Let this weight be w. Now, as it requires 80°C. of heat to convert a pound of ice at 0°C. into water at 0°C. (see Art. 117), the quantity of heat absorbed by the collected water will be expressed by $80 \times w$; whilst the quantity of heat given out by the body will be expressed by $W \times t \times x$, where x, as before, is the specific heat required. We have therefore the equation, $W \times t \times x = 80w$; hence $x = \dfrac{80\,w}{W\,t}.$

A certain amount of error results in the use of this instrument, from the fact that *all* the water does not escape; part of it adheres to the ice in its half-melted state.

113. Table of Specific Heats. — The following table gives the specific heats of certain bodies :—

MEAN OF SPECIFIC HEATS BETWEEN 0°C. AND 100°C.

Water............	1·0000	Iron	·1138
Alcohol..........	·4534	Copper.........	·0951
Mercury.........	·0333	Lead............	·0314

The above table gives the *mean* or *average* specific heats between 0° and 100°. It has been found that the specific heat of bodies increases with the temperature, and more so in liquids than in solids. In the case of water, however, this increase is less than in solids.

114. Experimental Illustration.—The difference which subsists between bodies, in regard to their capacity for heat, may be strikingly shown by the following experiment :—A cake of bees'-wax is placed upon the ring of a chemical stand (fig. 83). Three balls of different metals— iron, copper, lead, are immersed in a bath of very hot oil till they all acquire its temperature. If now they be taken out and put upon the cake, they make their way

through at different rates—the iron ball first, the copper
next, and last of all the lead.

**115. Influence of the High
Specific Heat of Water on
Climate.**—The high specific
heat of water plays an important part in the economy
of nature. The specific heat
of air has been found to be
nearly 4·2 times *less* than that
of water. It follows therefore
that 1 lb. of water in losing
1° C., would warm 4·2 lbs of
air 1° C. But water is 770
times as heavy as air; hence,
comparing equal volumes, a
cubic foot of water in losing
1°C., would raise 770 × 4·2, or 3234 cubic feet of air
1° C. We see from this, "the great influence which the
ocean must exert on the climate of a country. The heat
of summer is stored up in the ocean, and slowly given
out during the winter. Hence one cause of the absence
of extremes in an island climate."[*]

Fig. 83.

116. Latent Heat.—During the passage of a body from
the solid to the liquid state, or from the liquid to the
gaseous state, its temperature remains *constant*, whatever
be the intensity of the heating source. The heat which
the body receives in its *transition state*, does not affect
the thermometer, does not manifest itself, and on this
account it is called "latent." We may define *latent heat*,
therefore, as *the quantity of heat which disappears or is
lost to thermometric measurement, when the molecular constitution of a body is being changed.*

Thus if we take a block of ice, say at – 10° C., and apply
heat to it, its temperature rises till it comes up to 0° C.
At this point the temperature remains stationary until
the last particle of ice is melted. When this takes place,

* Tyndall on *Heat as a Mode of Motion*, p. 143.

the temperature again rises till it reaches 100° C., when it once more remains stationary, the water then gradually passing off in the form of steam.

117. Latent Heat of Water and Steam—(1) *Water.*— If 1 lb. of water at 80° C. be mixed with 1 lb. of water at 0°, the temperature of the mixture is 40° C. But if 1 lb. of water at 80° C. be mixed with 1 lb. of pounded ice at 0°, there will result 2 lbs. of water at 0° C. It follows therefore that 1 lb. of ice at 0° C., in being changed into 1 lb. of water at 0° C., requires as much heat as would raise 1 lb. of water through 80° C., or, which is the same thing, as would raise 80 lbs. of water 1° C. Consequently the number 80° C. (144° F.) expresses the latent heat of water or of the fusion of ice.

(2) *Steam.*—The latent heat of steam may be determined by observing the time required to raise a given quantity of water through a certain number of degrees, and then comparing this with the time between the commencement of boiling and the total evaporation of the water. It has been estimated at 540° C., implying that during the conversion of 1 lb. of water at 100° C. into 1 lb. of steam at the same temperature, as much heat is imparted as would raise 540 lbs. of water 1° C.

The latent heat of steam is of service in cookery. Vegetables and meat are often cooked by allowing the steam from boiling water to pass through them; in doing so, the steam becomes condensed and parts with its latent heat. We can easily understand from this the severity of a scald from steam.

Solution of Questions.—The student would do well to note the following questions, and the method of solving them :—

Ex. 1.—*How many pounds of ice at 0°C. can be melted by 1 lb. of steam at 100°C. ?*

Let *x* be the number. Then since 1 lb. of ice requires 80 units of heat to convert it into water, *x* lbs. will require 80 × *x*. Again, 1 lb. of steam at 100°C., in being converted into water at 100°C., *gives out* 540 *units of heat, and the* 1 *lb. of water has further to give out* 100 *units; hence the whole heat given out by the steam when reduced to water at* 0°C. = 540 + 100. *But this heat is absorbed by the ice, therefore we have* 80 × x = 540 + 100, *and* x = 8 *lbs.*—Ans.

Ex. 2.—*How many pounds of steam at* 100°C. *will just melt* 100 *lbs. of ice at* 0°C. ?

If x *be the number, then the quantity of heat given out by* x *lbs. of steam at* 100°, *when reduced to water at* 0° = 540 x + 100 x; *whilst the quantity of heat required by the* 100 *lbs. of ice to convert it into water at* 0° = 100 × 80; *hence* 540 x + 100 x = 100 × 80, *and* x = 12½.—Ans.

Ex. 3.—*What weight of steam at* 100° C. *would be required to raise* 500 *lbs. of water from* 0°C. *to* 10°C. ?

Let x *be the number of pounds. Here the quantity of heat given out* = 540 x + 90 x (*the water at* 100° C. *is to sink to* 10° C.); *hence* 540 x + 90 x = 500 × 10, *and* x = 7·9 *lbs.*—Ans.

Ex. 4.—*If* 4 *lbs. of steam at* 100°C. *be mixed with* 200 *lbs. of water at* 10°C., *what will be the temperature of the water ?*

Let x *be the temperature.* 4 *lbs. of steam in becoming water give out a quantity of heat* = 4 × 540, *and produce* 4 *lbs. of water at* 100°; *further, the* 4 *lbs. of water have to give out the additional heat* 4 (100 − x). *Again, the* 200 *lbs. of water in rising from* 10 *to* x, *absorb a quantity of heat* = 200 (x − 10); *hence* 4 × 540 + 4 (100 − x) = 200 (x − 10), *and* x = 22°·3.—Ans.

118. Cold of Evaporation.—In the passage of water or any other liquid into vapour, there is a quantity of heat rendered latent. This heat is chiefly derived from the liquid itself, hence the temperature of the liquid is lowered. We have therefore the important fact that *cold is produced by evaporation.* The more rapidly evaporation proceeds, the degree of cold is the greater. If, for example, we take the three liquids, water, alcohol, and sulphuric ether, and place a drop of each successively on the hand, then waving the hand backwards and forwards in the air to hasten the evaporation, we find that the sensation of cold is least with the water, greater with the alcohol, and still greater with the ether. This arises from the rate of evaporation at the same temperature being different in the three liquids.

119. Freezing by Evaporation. — Evaporation may proceed so rapidly as to cause refrigeration. This may be effected in the following manner :—A small capsule

Fig. 84.

containing water is placed in a flat dish filled with sulphuric acid. The whole is placed under the receiver of an air-pump, and the air exhausted. As the rarefaction proceeds the water evaporates, the vapour being immediately absorbed by the sulphuric acid, till the remaining water begins to freeze, and eventually becomes a solid lump. This experiment is due to Leslie.

Another experiment consists in filling a test tube with water, surrounding it with cotton wool saturated with sulphuric ether, and blowing a stream of air upon it by means of a pair of bellows (fig. 82). The evaporation of the ether takes place rapidly, and the water in a short time becomes frozen.

The method often followed out in India of procuring ice, affords an illustration of the same thing.

Early in the cold weather, when the nights are clear, shallow unglazed earthenware pans filled with water are put out in the open air. Evaporation rapidly takes place, and during the process, when the temperature falls below the freezing point, a thin stratum of ice forms on the surface of the water. Before daybreak the thin cakes of ice are removed from the pans, and the accumulated mass, well hammered together, is stowed away in the ice-house.

Water coolers, so much used in summer, owe their action to the same principle.

Fig. 85.

A remarkable instance of freezing by evaporation occurs in a grotto near Vergy in France. In some places columns of ice appear to support the vault of the grotto; at others they are seen hanging from the roof, or resting upon the ground. The water passes slowly in traversing the vault, and its evaporation hastened by currents of air produces the ice. It is not in winter alone that this takes place, nor can the formation of the "*glacières naturelles*" be attributed to a cooling down of the air.

120. The Cryophorus.—This instrument, invented by Wollaston, is founded on the same principle. It is represented in fig. 85. It consists of two glass bulbs, A and B, connected by a tube. Water is put in the bulb A, and whilst a small orifice is left open at the bottom of the bulb B, the water is boiled ; the steam escaping from the water chases out the air, and when this is all expelled, the orifice is closed by means of a blowpipe. On the water regaining its ordinary temperature, there is left in the apparatus nothing but a little water and its vapour. If now the bulb A be placed in a vessel to get rid of currents of air, whilst the other bulb B is plunged into a freezing mixture, such as snow and salt, the vapour as it escapes from the water is condensed, and in the course of half an hour or so the water in A begins to freeze.

CHAPTER VI.

121. Convection of Heat.—By the *convection* of heat is meant that process by which heat is *carried* and distributed through the mass of a fluid body by the actual motion of its own particles. Thus, water is boiled by convection. When heat is applied to a vessel of water, as in fig. 86, there are produced a series of ascending currents which carry the heat to the other parts of the liquid, until the water is raised to the boiling point. In like manner, the air at the top of a room is heated by the ascending currents of warm air. The phenomena of winds are due also to convection.

122. Conduction of Heat.—Heat is said to be *conducted* when it is propagated along the *molecules* *of a body.* There is a great difference between bodies in regard to their conducting power, or *conductivity*, as it is more generally called. Thus, if a rod of iron and a rod of wood of the same length and diameter be taken, and an end of each be inserted in the fire, in a short time the other end of the iron rod will become heated, whilst no trace of heat will yet appear at the other end of the wooden rod. In such an experiment, therefore, it is clear that the heat has found a ready passage through the iron rod, and has met with considerable resistance in the wooden rod. Hence we may divide bodies, considered in relation to their conductivity, into two classes : *good* and *bad* conductors. Under the former class may be included the metals, and under the latter such bodies as wood, stone, glass, straw, wool, cotton, silk, etc.

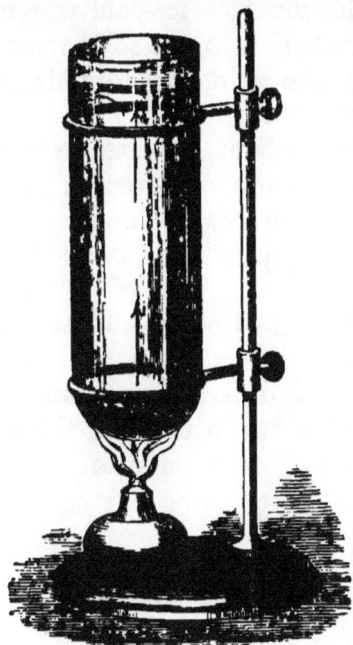

Fig. 86.

All liquids and gases possess a very feeble conducting power. If, for example, a vessel of water be heated *from the top* by pouring gently on the surface a quantity of boiling oil, it is found that the heat makes its way downwards with extreme slowness, and it is only after a considerable time that the least rise in temperature is observable at the bottom of the vessel.

Snow is a very imperfect conductor of heat. Travellers, when overtaken by a snow-storm, in some instances have had their lives preserved by taking shelter in a wreath of snow, before being benumbed by the cold. So also sheep have been taken out alive, though buried amidst snow for some time.

The Esquimaux, it is said, construct their winter huts of snow. They shape the snow into large hard masses, which they place upon each other as our masons do stones; they then pour into the crevices ice-cold water, which upon freezing unites the whole into one solid mass. The inside being covered with the skins of animals, a comfortable dwelling is thus provided.

This quality of snow is not without its use in the general economy of nature. In severe climates, it prevents the earth from being so much cooled down as to endanger those germs of vegetation which await the return of spring.

123. Relative Conductivity of Bodies. — Several methods have been followed out with a view to determine the relative conductivity of different substances. One method is this : A rectangular bar of uniform thickness (fig. 87) is heated at one end. A number of holes

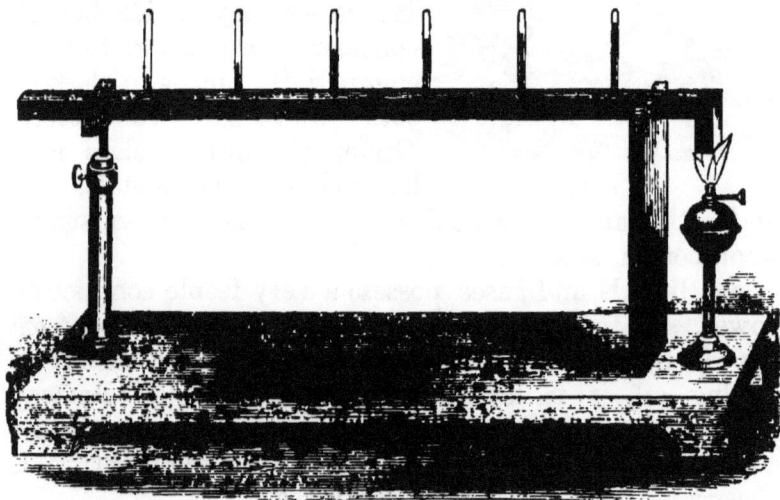

Fig. 87.

are made at equal distances along the bar, sufficient to hold the bulbs of so many thermometers. When the heating source is applied, the thermometers begin to rise, but at very different rates—the one nearest rising the fastest, whilst the one farthest away is but little affected. The rates at which the different thermometers rise are then carefully noted. The same thing is done with bars of different material; and by comparing the rates of ascent of one set of thermometers with those of others, the relative conductivities of the substances are ascertained.

The following table has been constructed from such investigations:—

Name of Substance.	Conductivity.
Silver,	100
Copper,	74
Gold,	53
Iron,	12
Lead,	9
Platinum,	8
Bismuth,	2

It is interesting to note the fact that the numbers contained in the above table *nearly* express the conductive powers of the bodies for electricity.

124. Experimental Illustration. — The difference in the conductivity of metals may be illustrated by the following simple experiment.

Two bars of different metals, such as copper and iron, are placed as in fig. 88. At equal distances along the

Fig. 88.

bars are attached a series of wooden balls by means of wax. When the ends are heated by a lamp, the heat is propagated along the bars, but as the copper is a better conductor than the iron, the wax on the former is more readily melted, and a greater number of balls fall off from the copper than from the iron in the same time.

125. Effect of Mechanical Texture.—Mechanical texture has an effect on the conduction of heat. Thus, twisted silk conducts heat more readily than raw silk; hard rock crystal more readily than when reduced to powder; wood more readily than in the state of sawdust. The reason is that in the latter cases the molecular chain is not so continuous; it is broken up by air-spaces.

126. Clothing.—As the object of clothing is to prevent the escape of heat from the body, we must of course select those substances as articles of dress which offer resistance to the passage of heat, or such as are bad conductors. The common notion that there is natural warmth in any material is quite a wrong one. There is really no more natural heat in a piece of flannel than there is in a piece of lead. Flannel is an excellent covering for a man in winter; it is nevertheless also the best substance for wrapping round ice to prevent it melting in summer. In the former case the source of heat being within, the flannel prevents the escape of heat, and thus contributes largely to warmth; in the latter case, the source of heat is from without, and the flannel being a bad conductor effectually prevents the passage of heat into the ice.

There being therefore no such thing as natural warmth in any material, it is evident that the lower the temperature to which we are exposed, the greater the waste of animal heat would be; hence in cold weather it becomes necessary to surround the body with such materials as are the worst conductors of heat. Now, according to experiment, fur is the worst conductor, and therefore the warmest covering; next to it is wool, fabricated into the different textures of flannel and cloth; next are cotton, linen, and silk, which, being better conductors, form therefore a comparatively cool covering, and are fit only for the higher temperatures of summer.

Air is a bad conductor of heat; hence loose clothing is warmer than we are apt to imagine.

127. Sensations of Heat and Cold.—The different sensations of heat and cold, which we continually experi-

ence in *touching* bodies, arise altogether from conduction.
When two bodies of different temperatures are placed in
contact, the warmer parts with its heat to the colder,
until they both acquire the same temperature. There is a
constant tendency towards an *equilibrium of temperature.*
Suppose, then, that a person in a room without a fire
were to touch first the carpet, then the table, then the
wall, and lastly the fender, he would consider each of
them colder and colder in succession. Why? The reason
is simply this: the carpet being a bad conductor, carries
little heat off from the hand; the table is a better conductor, and thus feels colder; the wall is a better conductor still, and therefore feels still colder; but the fender
is the best conductor of the whole, and accordingly it
carries off the heat rapidly, giving thereby the most powerful sensation.

128. Combustion. — Combustion, such as we have it
in our coal, in our gas and candle flames, is due to the
chemical union of the oxygen of the air with the substances present in these.

Coal-gas is a chemical combination of carbon and
hydrogen. When the jet of escaping gas is ignited,
"the oxygen of the air unites with the hydrogen,
and sets the carbon free. Innumerable solid particles of carbon thus scattered in the midst of the
burning hydrogen, are raised to a state of intense incan-
descence: they become white hot, and mainly to them
the *light* of our lamps is due. The carbon, however, in
due time, closes with the oxygen, and becomes, or ought to
become, carbonic acid; but in passing from the hydrogen,
with which it was first combined, to the oxygen with
which it enters into final union, it exists for a time in
the solid state, and then gives us the splendour of its
light." Within the flame there is a core of unburnt gas.

"The combustion of a *candle* is the same in principle
as that of a jet of gas. On igniting the wick, it burns,
melts the tallow at its base, the liquid ascends through
the wick by capillary attraction, it is converted by the

heat into vapour, and this vapour is a hydro-carbon, which burns exactly like the gas."*

129. Structure of a Candle Flame.—It is to Sir Humphry Davy that we owe our knowledge of the precise theory and constitution of flame. The structure of a candle flame will be understood from fig. 89. It consists of three parts: (1) the space occupied by the unburnt vapour; (2) the luminous zone or area where the carbon particles are in a white-hot, glowing state; (3) the area of complete combustion, from which the greatest amount of heat is evolved. The presence of unburnt vapour within may be shown by placing a small glass tube, as in the figure. The vapour escapes through the tube, and may be ignited at the other end.

The same thing may be shown by lowering a piece of white paper upon the flame till it nearly touches the

Fig. 89.

wick. A blackened or charred ring is formed upon the paper, whilst within the ring the paper is unaffected.

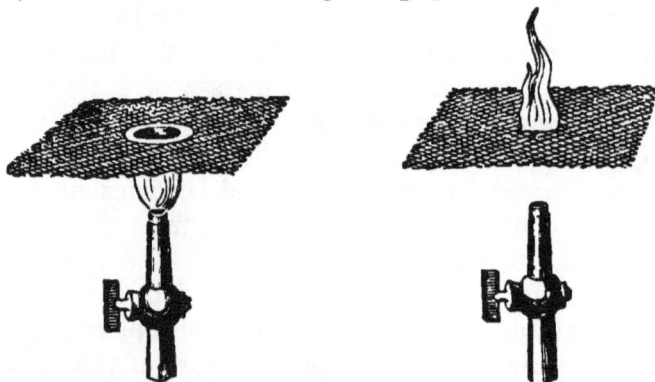

Fig. 90.

130. Effect of Wire Gauze.—If a piece of fine wire

* Tyndall on *Heat as a Mode of Motion*, pp. 46, 47.

gauze be lowered upon a gas jet the flame spreads out below (fig. 90), but is unable to penetrate the meshes of the gauze. This is owing to the conduction of the heat by the gauze, in consequence of which the gas that escapes through cannot become ignited. If, whilst the gas is escaping, the gauze be held a little above the burner, it may be ignited from above, but the flame cannot reach the gas below; and it may be extinguished by raising the gauze quickly upwards.

On this principle is constructed the "Davy Safety Lamp," so much used by miners.

131. Bunsen Lamp.—The luminosity of flames, as we have seen, is mainly due to the existence of solid carbon particles. Hence when a large quantity of air is allowed to mix with them their combustion is quickened, and heat is developed at the expense of intensity of light. This is what is effected by a *Bunsen lamp,* so much used in chemical and physical laboratories. It is represented in

Fig. 91.

fig. 91. The gas, escaping from a central burner, up the tube, draws with it a quantity of air through the small holes near the base. The mixture of gas and air is then ignited at the top of the tube, and burns with a feeble light, but evolves considerable heat, owing to the complete combustion of the carbon. If the small holes be closed, the flame assumes its ordinary appearance.

132. Animal Heat.—The heat of our bodies is due to a slow combustion constantly going on. The oxygen of the air we inspire combines with the carbon elements of the blood and animal tissue, and by their union heat is evolved—the carbonic acid thus formed being constantly exhaled. The air we expire contains from 3 to 6 per cent. of carbonic acid, and will not support the combustion of a candle.

CHAPTER VII.

133. Radiation of Heat—Theory of Exchanges.— There is no quality so abundantly obvious in reference to heat, than its tendency to diffuse itself in all directions from the heating source. This passage of heat through intervening space is called *radiation*, and the heat thus passing, *radiant* heat.

Radiation is not dependent upon the presence of air; it takes place also in a vacuum. This is manifest when we consider how it is that we derive heat from the sun; his heating rays require to pass through an intervening void before they reach our earth.

We are accustomed to speak of warm bodies only radiating heat; but the fact is that *all* bodies, of whatever temperature, radiate heat. Let us suppose we have two bodies, A and B, of different temperatures, A warmer than B. Radiation takes place not only from A to B, but also from B to A. However, in consequence of A's excess of temperature, more heat passes from A to B than from B to A, and this continues until both bodies acquire the same temperature.

At this point the radiation does not cease; but now the amount of radiation is the same for both—as much heat passes from B to A as from A to B, or the one body gives out as much heat as it receives from the other. This theory is known as "Prevost's Theory of Exchanges."

If we place ourselves near a block of ice, we experience the sensation of cold. This might lead us to the belief that cold is a separate influence, and can be radiated like heat. In this case, however, the body being warmer than the ice, there is a greater radiation from it towards the ice than from the ice towards it, hence the cause of the sensation.

134. Reflection of Radiant Heat.—Heat, like light, is capable of reflection, and follows the same law (Art.36).

The reflection of heat is well illustrated by the apparatus represented in fig. 92. Two metallic reflectors mounted on stands are set directly opposite each other. A white-hot

Fig. 92.

iron ball is placed in the principal focus of one of the reflectors; if now a piece of phosphorus be placed in the focus of the other reflector, it will burn, being fired by the heat emitted from the ball, which has been concentrated by the reflectors at that point.

The reflective powers of substances vary considerably. According to Leslie's experiments, the greatest reflection takes place from bright and polished metallic surfaces. Should the surface be rough or tarnished, the amount of reflection is much diminished. Glass coated with lampblack, and white paper, reflect very feebly.

135. **Radiating Power of Bodies.**—Experiment shows a marked difference also in the powers of bodies to radiate heat. A body which reflects heat well is found to

radiate badly ; in other words, a good reflector of heat is a bad radiator, and *vice versâ.* The two qualities, in fact, of reflection and radiation, are directly opposed to each other.

In the interesting researches on this subject by Leslie, he made use of a tin canister mounted on a stand and filled with hot water, the sides of which were coated over with different substances, *e.g.,* one with lampblack, a second with writing paper, a third with glass, and a fourth with a layer of silver. He found the radiating powers of the faces to be very different. Expressing the radiation of lampblack by 100, he found that of the paper to be 98, of the glass 90, and of the silvered surface 12. This apparatus is known as *Leslie's cube* or *canister.*

The high radiating power of fire-clay is well known, hence the common expedient of lining a grate with this substance, so as to increase the radiation from the fire. In regard to the metals, it may be stated generally, that the brighter and more polished the surface, the less the radiation. Of all substances, lampblack possesses the highest radiating power.

136. Strange Effect of Close Contact.—Under certain circumstances the cooling of a vessel containing hot water may even be hastened by surrounding it with flannel. Thus, if two similar vessels be taken, both filled with hot water—the one *closely* enveloped in flannel, and the other left uncovered—the former is found to radiate more freely, and after a time to become sensibly cooler than the other.

137. Application to Common Experience.—We may gather from the foregoing principles many useful and important hints regarding facts of every-day life. Thus we learn why the polished fire-irons, which stand beside a fire, are not inconveniently heated. The heat which falls upon them is reflected in a great measure by the polished metal. Should they be allowed to become tarnished the reflection is not so complete, and they become heated. The polish, therefore, of fire-irons is not only ornamental, but contributes largely to comfort in handling them. It is of advantage that the interior of a screen placed behind roasting meat be kept clean and polished, for then it is a good reflector, and aids materially the cooking process.

Certain parts of a steam engine ought to be highly polished, not so much for appearance' sake, but as a most effectual means of retaining the heat of the steam, thus preventing loss by condensation. A stove ought to have its exterior surface rough and well blackened, so as to allow radiation to take place freely. A tea-kettle, on the other hand, ought to be well brightened up so as to diminish radiation, and thus tend to retain the heat of the water as long as possible. Should a "cosy" be used for a tea-pot, it ought to be made to fit loosely, for then the radiation is much impeded.

138. Absorbing Power of Bodies — Reciprocity of Radiation and Absorption.—By the *absorbing power* of a body for heat is meant that quality, in virtue of which it allows heat to pass into its mass. The heat which falls upon a body is in part absorbed and in part reflected. All the reflected portion, however, is not reflected regularly, that is, it does not follow the ordinary law of reflection (Art. 36); part of it, as in the case of light, is irregularly reflected, and follows no particular law—it is called *scattered* or *diffused* heat. We may infer from this that a body which is a good reflector of heat is a bad absorber, and *vice versâ*. This has been corroborated by experiment.

Again, it has been found that the two qualities of radiation and absorption are *reciprocal*, that is, a body which is a good radiator of heat is also a good absorber, and one which is a bad radiator is also a bad absorber.

To ascertain the relative absorptive powers of different kinds of cloth, Dr. Franklin made the simple experiment of putting a number of pieces on snow as it lay on the ground. These were exposed for a certain time to the sun's rays, and the depths to which they severally sank in that time were noted. He found those pieces that were dark in colour sank deepest in the snow, while those that were light-coloured sank least, from which he inferred that the former possessed the greatest power of absorption, and the latter the least. Hence appears the importance of attending to the particular colour of clothing which should be worn in the different seasons. Thus, the worst colour of cloth we can wear in winter is black; for, being a powerful radiator, it tends to carry off the heat from the body. In summer, again, a light-coloured dress is the most desirable; for, being a good reflector and a bad absorber, it shields the body from the influence of the sun.

The discovery ships of Captain Parry, it is said, during the severe winter which was spent at Melville Island, were so rigidly frozen in as to render it extremely doubtful whether the influence of the summer's sun would be sufficient to relieve them. To ensure an exit, the method was adopted of strewing ashes and soot in a line from the ships to seaward. The consequence was that these substances, by their great absorption of the sun's rays, dissolved the subjacent ice, thus forming a passage for the ships through the solid ice all around.

In some of the more mountainous districts of Europe, where the snow would lie so long as to retard cultivation, the peasantry have recourse to the plan of strewing a quantity of earth upon the snow; this, by its great absorptive power, assists materially towards clearing the ground.

139. Refraction of Heat.—That heat, like light, can be refracted, is plain from the simple expedient of concentrating the sun's rays by a burning-glass. Experiment proves that a beam of radiant heat is made up of rays of different degrees of refrangibility. Most sources of heat emit heat rays, which are partly *luminous* and partly *obscure*, and those differ from each other in regard to their refractive capabilities.

A bottle of water, acting like a lens, has been known to converge the sun's rays to such an extent as to cause conflagration. So also, it is said, that in greenhouses drops of water on the plants sometimes exercise such convergence on the solar beams, as to burn up the leaves.

140. Diathermancy. — By this term is meant the *power of a body to transmit radiant heat.* It bears the same relation to radiant heat that transparency does to light. A body, however, which is transparent to light, does not necessarily possess diathermancy. Thus a sheet of ice, though transparent, does not transmit much heat.

The diathermic power varies much in different bodies. The following table exhibits some of the results obtained by Melloni, whose researches on this subject have been very extensive :—

SOLIDS.

Name of substance. (Thickness = $\frac{1}{10}$th of an inch).	Transmission. (Percentage of the total radiation).
Rock-salt	92·3
Fluor-spar	72
Glass	39
Felspar	23
Alum	9
Ice	6

LIQUIDS.
(Thickness = ·36 in.)

Bisulphide of carbon	63
Sulphuric acid	17
Distilled water	11

It appears from this table that rock-salt has a high diathermic power, about $2\frac{1}{3}$ times that of glass, 10 times that of alum, and 15 times that of ice. Melloni has found that the power of transmission varies in different bodies with the *source* of heat—rock-salt, however, forming an exception.

The two substances, rock-salt and alum, are used to separate the light and heat which radiate from the same source. The former, when covered with lampblack, transmits the heat freely, but arrests the light; whilst the latter arrests the heat and transmits the light.

Aqueous vapour, though diathermic for heat from luminous rays, is not so for heat from obscure rays.

Tyndall has proved this by a series of careful experiments. He remarks in regard to it, "No doubt can exist of the extraordinary opacity of this substance to the rays of obscure heat; particularly such rays as are emitted by the earth after being warmed by the sun. Aqueous vapour is a blanket more necessary to the vegetable life of England than clothing is to man. Remove for a single summer night the aqueous vapour from the air which overspreads this country, and you will assuredly destroy every plant capable of being destroyed by a freezing temperature. The warmth of our fields and gardens would pour itself unrequited into space, and the sun would rise upon an island held fast in the iron grip of frost."* He has also shown that air, more or less charged with aqueous vapour, may exercise from 30 to 70 times the *absorptive* effect of dry air. Hence appears the cause of the extreme cold met with in the upper regions of the atmosphere, where the quantity of aqueous vapour is much reduced.

It is worthy of note that glass is capable of transmitting *luminous* heat, but greatly retards the passage of *obscure* heat. As the sun's radiation consists in a large measure of luminous heat-rays, we can understand why the panes of glass in a window are not much heated even by brilliant sunshine. By their contact with different objects, however, they are changed into obscure rays, and as such cannot re-traverse the glass. Hence the reason why a room exposed to a summer's sun gets so heated— the glass, though allowing the sun's heat to pass in, serves as a barrier to its getting out. Hence also the high temperature of green-houses and photographic apartments after strong sunshine.

The effect of a glass screen placed in front of a fire is well known. The calorific rays being in a large measure intercepted, the screen becomes warm, but radiates its heat in all directions, and thus the heat of the fire is mitigated, though at the same time we have its pleasant light.

* Tyndall on *Heat as a Mode of Motion*, p. 372.

QUESTIONS.

1. What is meant by (1) the *linear*, and (2) the *cubical co-efficient* of expansion? Is there any relation between them? if so, state what it is.

2. Convert $-15°$C. into the Fahrenheit scale; and $12°$R. into the centigrade.   *Ans.* (1) $5°$, (2) $15°$.

3. At what temperature is the density of water a *maximum?* Water expands in freezing: is this property peculiar to water?

4. Explain the process by which a lake is frozen over.

5. Explain the trade-winds.

6. Define "specific heat." Describe some method of determining the specific heat of a body.

7. A pound of mercury at $102°$ is immersed in a pound of water at $40°$; how much will the temperature of the water be raised, assuming the specific heat of mercury to be $·03$?
       Ans. $2°$.

8. What is meant by "latent heat?" What is the latent heat of water? Explain clearly what the number implies.

9. How many pounds of ice at $0°$C. can be melted by 5 pounds of steam at $100°$C.?   *Ans.* 40.

10. If 1 pound of steam at $100°$C. be mixed with 49 pounds of water at $15°$C., how much will the temperature of the water be raised?   *Ans.* $12\frac{1}{2}°$.

11. Give Prevost's *theory of exchanges.*

12. Describe some experiment for freezing water by evaporation.

13. Distinguish between the *convection* and *conduction* of heat.

14. Give the theory of combustion of a gas flame; and explain the action of a Bunsen lamp.

15. The rays of the sun in passing through a window do not sensibly heat the panes of glass; but a glass screen placed in front of a fire becomes warm. How do you account for these effects?

SELECTION OF QUESTIONS

PROPOSED AT THE

EXAMINATIONS OF THE GOVERNMENT DEPARTMENT OF SCIENCE AND ART, FROM 1867 TO 1872, WITH THEIR SOLUTIONS.

ACOUSTICS.

1. *What do you understand by a wave of sound?*

By a wave of sound is meant an undulation, or wave-like motion, communicated to the particles of air in consequence of the vibration of the sounding body. It consists of two parts, one in which the air is condensed, called a *condensation*, and the other in which the air is rarefied, called a *rarefaction*.

Whilst the undulation or wave is transmitted through the air, the aerial particles themselves have but a very small motion to and fro, in the direction in which the sound is propagated. Such oscillatory movements are necessarily attended by condensations and rarefactions.

2. *Describe the manner in which sound is propagated through air, water, or wood.*

When a body in air emits sound, it moulds the surrounding air into undulations or waves. These waves consist of alternate condensations and rarefactions of the air caused by the vibration of the sounding body, and it is in consequence of such waves entering our ears that we derive the sensation of sound.

A similar effect is believed to take place in water or wood. The sound is propagated through either substance in the form of *waves*, in which the molecules have a limited motion to and fro.

3. *The velocity of sound in water is much greater than its velocity in air. Why is this the case?*

The *velocity* of sound in any medium depends upon the elasticity and density of the medium, or, rather, upon the *relation* which the former bears to the latter. Now, in the case of water, this relation is higher than in air, and hence the velocity of sound is greater.

8 E

I

4. You fire a shot before a cliff, and hear the echo five seconds afterwards, what is the cliff's distance from the centre of explosion ?

Sound travels at the rate of 1125 feet per second (temperature = 62°F.). In five seconds it will therefore travel over 1125 × 5, or 5625 feet, that is, in this time the sound of the explosion in going to, and returning from, the cliff, will travel over this space ; hence the distance of the cliff = $\frac{5625}{2}$ = 2812½ feet—Answer.

5. How do you suppose the human voice to be produced ? What occurs in the case of your voice when you sing high notes and low notes ?

The trachea, or wind-pipe, tapers towards the top, leaving a narrow slit-like opening between the membranes, termed the *glottis.* The air passing from the lungs with sufficient force, in escaping through the glottis throws the membranes into vibration, and thus it is that voice is produced.

Notes, or sounds of different pitch, are produced by the muscles, or *vocal chords,* as they are called, acting upon the membranes enclosing the glottis, either in the way of tightening or relaxing them, and this is effected by the simple act of volition on our part.

6. Air and hydrogen gas are urged in succession through the same organ-pipe. Describe the effects and explain them.

Air, when urged through an organ-pipe, gives a note of a certain pitch. When hydrogen gas is urged through the pipe with the same force it produces a note of a *higher* pitch. This results from the velocity of sound being greater in hydrogen than in air, and hence the number of vibrations per second executed in the former case is greater than in the latter.

7. How are musical sounds produced ? On what do the pitch and the intensity of a musical sound depend?

A musical sound is produced by periodic impulses imparted to the air by the sounding body, that is, by impulses which succeed each other after perfectly equal intervals of time.

Pitch depends entirely upon the number of vibrations executed per second.

Intensity, again, depends upon the amplitude of the vibrations, or the amount of disturbance given to the air-particles in consequence of these vibrations.

8. What are the changes of temperature which occur in a wave of sound ?

A wave of sound consists of two parts, in one of which the air is condensed, and in the other rarefied. Now, when air is condensed heat is evolved, and when rarefied cold is produced. In the condensed portion of the wave, therefore, the air is above, and in the rarefied portion below, its average temperature. These changes of temperature, however, though augmenting the velocity of sound, do not affect the general temperature of the air through which the sound passes.

9. *State the likeness that exists between sound and light as regards reflection and refraction.*

(1) *Reflection.*—Sound and light obey the same law, viz., the angle of reflection is equal to the angle of incidence. Their similarity in this respect is proved in several ways. Thus, in an elliptical whispering gallery, whilst the exhibitor places himself in one focus, the observer, by going to the other, can hear the slightest whisper. So also if two large elliptical metallic reflectors be so placed as to have their foci in the same position as they would be were the ellipse (or rather ellipsoidal shell) complete, a lamp set in one focus would have its light concentrated in the other, and an image of the lamp would be formed there.

Or, again, by taking a concave spherical reflector, and placing a watch immediately in front, its ticking may be heard distinctly by a person adjusting his ear at the focus some distance off. A lamp, in like manner, placed in the same position, will have an image of itself formed at precisely the same point as that at which the ticking of the watch was heard.

(2) *Refraction.*—Sound and light can also be refracted or bent out of their original course. A common method of refracting light is to use a glass lens. In the case of sound, it can be refracted by using a thin india-rubber balloon filled with carbonic acid gas. A watch placed at one side of it can be distinctly heard at the other, near the point where the sound-waves are converged.

10. *Wherein does the transmission of sound through a smooth tube differ from its transmission through the open air ?*

The sound-waves which proceed from a sonorous body gradually enlarge as they recede from it, in open air. When they are transmitted through a smooth tube, they are prevented from thus enlarging, and, in addition, are reflected from side to side of the tube till they eventually emerge at the end. In consequence of this, the slightest whisper can be heard through a long tube.

11. *What is the influence of heat and cold upon the velocity of sound in air ?*

If the *density* of the air be diminished, the velocity of sound is increased, and if the density be increased, the velocity is diminished (the elasticity remaining the same). Now the effect of heat is to diminish the density of the air, and of cold to increase it. Hence heat increases the velocity of sound, and cold diminishes it.

LIGHT.

1. *A straight stick placed obliquely in the water appears bent at the surface of the water, a ray of light entering the water obliquely is also bent at the surface; are the stick and the beam bent in the same manner or not ?*

No. The stick is bent upwards, or from the perpendicular, whilst a ray of light, entering the water obliquely, is bent *towards* the perpendicular; because in the latter case the ray passes from a rare medium into a denser.

2. *Describe an experiment to prove that out of a mixture of light of various colours you can produce colourless or white light.*

Take a circular disc of cardboard and divide it into sectors, having the same proportion as the spaces which the different colours of the solar spectrum occupy. Then colour these sectors with the different hues successively, viz., red, orange, yellow, green, blue, indigo, and violet.

If now the disc be attached to a revolving apparatus, and whirled rapidly round, the colours will be blended together, and will produce a white appearance. The experiment is founded on the principle of the "persistence of impressions on the retina."

3. *State what you know regarding the form and use of spectacles.*

(1) *Form.*—Spectacle glasses are either converging or diverging lenses. Of each class there are three kinds, which have names assigned them from the nature of their bounding surfaces. A *double convex* lens is a type of the converging class, and a *double concave* lens of the diverging class.

(2) *Use.*—In order to have distinct vision, the image of an object must be thrown *upon the retina.* The object of spectacles is to enable the eye to accomplish this. In far-sighted persons, the eye has too little convergent power, and cannot concentrate the rays of light upon the retina. On the other hand, in short-sighted persons, the eye has too much convergent power, and brings the rays to a focus too soon, or in front of the retina. The remedy in the former case is to use a converging glass, and in the latter a diverging glass, of sufficient strength to enable the eye to concentrate the rays exactly upon the retina.

4. *State what you know regarding the production of the colours of flowers. Why, for example, is a rose red and grass green ?*

The colour of an object depends upon the treatment to which the rays of light falling upon it are subjected. Sun-light is not homogeneous, but is composed of seven distinct kinds of light. Colour is owing to certain of these constituents being absorbed or quenched by the object, the remaining ones being reflected and giving to the object the colour which it appears to possess.

Thus, a rose is red because all the constituents of white light, except the red (which is reflected), are absorbed; so also the grass is green, because the green colour is reflected by the grass, the other constituents being absorbed.

5. *A cloud is composed of transparent water-particles; but, if transparent, why are clouds able to intercept so much of the sun's light?*

The water-particles, though transparent, are not perfectly so; part of the light falling upon them is absorbed, but the greater part is broken up and scattered in all directions by multiplied reflections. There is consequently a loss of light, and this loss will be greater the denser the cloud.

It is for the same reason that a mass of pounded glass becomes practically opaque when in sufficient thickness.

6. *I wish you to compare the light of a candle flame with that of a gas flame of the same size. How would you determine and express, numerically, the relative intensities of the two lights?*

The relative intensities of two different sources of light may be determined by what is called "the shadow test." The method is this: In a darkened room place the two lights in front of a screen, and between them and the screen place a rod or stick; now adjust the lights at such distances as that the shadow of the stick may appear equally illuminated when viewed from the other side of the screen. Then measure these distances. Suppose they are for the candle and gas flame 5 feet and 8 feet respectively. Now, as the intensities are directly proportional to the squares of the distances, we have .

Intensity of candle : intensity of gas : : 25 : 64;

in other words, the gas would have nearly $2\frac{3}{5}$ times the illuminating effect of the candle.

7. *A large concave mirror is placed before you. You see your image first inverted in the air; you change your distance from the mirror, and find that in a certain position your image vanishes; again you change your position, and find your image erect. Under what circumstances are these effects observed? State whether the images observed are of greater or less size than yourself, and give the reason of the increase or diminution.*

When the image of the object is seen inverted in the air, the object itself is *beyond* the centre of curvature of the mirror. The image (real) in this case is formed between the principal focus and the centre of curvature, and is smaller than the object, because its distance from the centre of curvature is less than the distance of the object from the same point.

When the image vanishes, the object coincides with the centre of curvature, for in that position the rays from the object are reflected directly back, and the place of the object is the focus of the image.

Lastly, when the image (now virtual) is seen erect, the object is placed between the principal focus and the mirror. This image is larger than the object, because its distance from the centre of curvature is greater than the distance of the object.

8. *Describe the circumstances under which total reflection takes place. Supposing your eyes were placed under the water of a lake, what appearance do you suppose a man standing on the brink of the lake would present to you?*

When a ray of light passes from a dense medium into a rarer, it is refracted *from* the perpendicular.

If the angle of incidence be gradually increased, it will at last attain such a magnitude as that the emergent ray becomes nearly parallel to the surface of the water. This angle is called the *limiting* or *critical* angle of refraction. For water and air this angle is 48½°. If, therefore, the incident ray make a greater angle than this, it will not emerge from the water, but will be reflected at the surface, following the ordinary law of reflection.

As a ray of light passing from air into water is refracted *towards* the perpendicular, an eye receiving that ray will take it as coming along the line of prolongation of the refracted ray; hence, on the supposition stated, the man, as well as the brink of the lake, will appear lifted up.

9. *What is meant by the scattering of light, and what by its regular reflection?*

The scattering of light means the irregular reflection of light from the surfaces of bodies, in consequence of which we see them or become aware of their existence.

By the regular reflection of light is meant such reflection as we have from mirrors or polished reflectors; and, where the law holds good, that the angle of reflection is equal to the angle of incidence.

If a mirror reflected *all* the light which fell upon it regularly, the mirror itself would be invisible; it is in consequence of the scattering of some of the light that we are enabled to see it.

10. *I fill two cups of the same depth with two different liquids, and notice two things: firstly, both cups appear shallower than when they are empty; and, secondly, one of them appears shallower than the other. Explain the observation.*

The rays from the bottom of each cup, in emerging from the liquid, are refracted from the perpendicular, or towards the surface of the liquid. These rays therefore enter the eye as if they came along the prolongations of the lines in which they are refracted. The eye receives these rays as if they came from points above the bottom of each cup; hence each bottom appears raised; in other words, both cups appear shallower than they really are.

That cup appears shallower which contains the denser liquid, because the rays are subjected to a greater amount of refraction.

11. *Describe clearly an experiment by which white light can be resolved into the differently coloured lights which compose the white. Describe also an experiment by which the colours can be recompounded.*

A beam of sun-light is admitted through an aperture made in the shutter of a darkened room; a prism is interposed in its course, and a screen is placed at some distance from it. An elongated image of the sun is found depicted on the screen, and coloured after the following manner (commencing with the lowest colour): red, orange, yellow, green, blue, indigo, and violet. The image thus formed is called the solar spectrum.

For the answer to the remaining part, see Question 2.

12. *State the likeness that exists between light and radiant heat as regards reflection, refraction, and transmission.*

(1) *Reflection.*—Both obey the same law, viz., that the angle of reflection is equal to the angle of incidence. This is well proved by placing two concave spherical reflectors directly opposite each other, and placing a lamp or red-hot iron ball in the principal focus of one of the reflectors. It is found that the light or heat is concentrated in the principal focus of the other reflector, owing to the reflection from the two mirrors.

(2) *Refraction.*—A convex lens of glass is a common means of refracting light, so as to converge it to a focus. Radiant heat may also be concentrated by a lens of this kind, but better still by a lens made of rock-salt.

(3) *Transmission.*—Light is transmitted through a plate of glass, so also is heat, but not with such readiness as through a plate of *rock-salt*. This latter substance, whilst it offers resistance to the transmission of light, gives a ready passage to heat.

13. *Light issues from a luminous globe twelve inches in diameter, and falls upon a second opaque globe six inches in diameter, show by a diagram the kind of shadow cast by the latter.*

By examining carefully the diagram in Art. 30, the student will have no difficulty in constructing the diagram for this question.

N.B.—There are both the umbra and the penumbra formed.

14. *Looking through a red glass at the white body of the sun you see it red, in what way does the red glass act upon the light so as to produce this impression?*

The sun's light in traversing the glass is decomposed, all the constituents being quenched, except the red light; hence the red appearance given to the sun.

HEAT.

1. *What do you understand by radiant heat ?*

Radiant heat is the heat emitted by a body through intervening space. The heat of the sun, the heat of a fire, the heat of a stove, etc., are examples.

The heat of a body is believed to be owing to a state of vibration among its particles. This vibratory motion a heated body can communicate to the surrounding ether, which in turn affects other bodies, and thus heat is said to be capable of radiating from one body to another.

2. *How is the bulk of a body usually affected by heat ? Are there any exceptions to the general rule ? If so, state such exceptions, and describe the circumstances under which the exception appears.*

A body is enlarged in bulk by the addition of heat, and diminished by the abstraction of heat. The general rule may be shortly stated thus : "Heat expands, and cold contracts."

There are some exceptions to this principle. Water when cooled down to about 40° F. contracts, so far obeying the rule; at this point, however, all contraction ceases. When cooled down below this temperature it expands, and it does so at an increasing rate as the freezing-point is approached.

Certain metals, when melted, undergo expansion on solidification—bismuth and cast-iron are examples; hence the precision with which cast-iron takes the impression of a mould.

Another exception is found in stretched india-rubber. Thus, if an india-rubber band support a weight, and the position of the weight be observed, when the band is heated the weight is raised, owing to contraction on the part of the rubber.

3. *If a liquid be heated at the bottom, how is the heat distributed through the liquid? If heated at the top, how is the heat propagated?*

When a liquid is heated at the bottom, the heat is distributed through the liquid by *convection*. The particles of the liquid next the heating source becoming warm, expand and rise to the surface. Other particles take their place, which in turn also expand and rise. There is thus a circulation maintained in the liquid, in virtue of which the warm particles just escaped from the bottom are constantly ascending, whilst the colder particles are as constantly descending to supply their place.

If the liquid be heated at the top the heat is found to make its way very slowly downwards, and this it can only do by the heat being propagated from particle to particle, or by the process of *conduction*.

4. What is meant by the boiling point of a liquid? What is the boiling point of water, and how could water be heated above its ordinary boiling point ? How is the boiling point affected when we ascend a mountain ?

The boiling point of a liquid may be defined as "that point at which the tension or elastic force of its vapour is equal to the pressure which it supports."

The boiling point of water (when the barometer is 30 inches) is 212°F. At this temperature, according to the above definition, the elastic force of the vapour, or steam, which escapes from the water, is equal to the ordinary pressure of the atmosphere. By subjecting the water to a greater pressure than the ordinary pressure of the atmosphere, the boiling point of water could be raised; in other words, the water could be heated above its ordinary boiling point. This actually takes place in a closed vessel, such as in the boiler of a steam engine. The steam accumulates on the surface of the water, and exerts a greater pressure than that of the atmosphere; the consequence is that the water is heated above 212°F.

The boiling point of a liquid is lowered as we ascend a mountain, because of the diminishing pressure of the atmosphere. At the top of a high mountain, therefore, it may be lowered many degrees. At the top of Mont Blanc, for example, water is found to boil at 180°F.

5. Give a clear statement of what you understand by the radiation, reflection, and absorption of heat.

Radiation is the passage of heat from one body to another through intervening space, or it may be stated to be "the communication of motion from the particles of a heated body, to the ether in which these particles are immersed, and through which the motion is propagated."

Reflection is the bending back of radiant heat into the medium through which it came to meet the reflector. Radiant heat can be reflected from bright metallic reflectors, whether plane or curved, and follows the ordinary law of the reflection of light.

Absorption is the receiving or taking in of heat by a body from a warmer one. The two qualities of radiation and absorption are reciprocal; in other words, the radiating power of a body is just in proportion to its absorbing power.

6. I place water, alcohol, and ether, all of the same temperature, on my hand in succession. I experience a certain cold from the water, a greater cold from the alcohol, and a still greater cold from the ether. State the cause of these differences.

It is the rate of evaporation. When a liquid is being evaporated heat is abstracted, which becomes latent in the vapour, and the amount of that heat varies with the rate of evaporation of the liquid. In the question, the evaporation being aided by the heat of the hand, there is an abstraction of heat from the hand, producing thereby a cold sensation in each case. But the water evaporates slowly, the alcohol more rapidly, and the ether more rapidly still ; hence the difference in the sensations.

7. *A pound of iron at a temperature of* 100° *is immersed in a pound of water at a temperature of* 50°; *how many degrees will the temperature of the water be exalted?* (*Note.—Specific heat of iron* = ·1).

There is here a difference of temperature between the iron and water of 50°. But the specific heat of water is 10 times that of iron ; hence we have to divide 50° in the proportion of 10 to 1.

∴ 11 : 1 : : 50 = $4\frac{6}{11}$, that is, the water is raised $4\frac{6}{11}$°.

8. *A plate of rock-salt if placed in front of a fire will not be heated, while a plate of glass will be heated. A hot plate of rock-salt held at a short distance from the face hardly warms the face, while a hot plate of glass does warm it. Explain these effects.*

Some bodies are diathermic, that is, allow heat to pass through them freely; others offer considerable resistance. Rock-salt is an instance of the former class, and glass of the latter. The heat of the fire therefore is transmitted readily through the plate of rock-salt, whilst it is largely arrested by the plate of glass; hence the difference of effect.

Again, rock-salt is a bad radiator of heat, and glass is a good radiator. The rock-salt plate therefore radiates little heat towards the face, whilst the glass plate radiates pretty freely.

9. *How is the heat of a fire produced? How is the heat of your own body produced?*

The heat of a fire is produced by combustion, that is, by the chemical combination of the oxygen of the air with the carbon of the fuel. Ordinary coal consists mainly of carbon and hydrogen ; when it is lighted the oxygen of the air unites with the carbon and hydrogen, forming carbonic-oxide, carbonic acid, and water vapour, at the same time evolving heat.

The heat of the body is also due to combustion; the oxygen of the air unites with the carbon elements of the blood and animal tissue, evolving heat, and forms carbonic acid, which is being constantly exhaled.

10. *Are clothes really warm? If not, why are they sometimes called warm? What is the real meaning and action of cool dresses and warm dresses?*

No. They are popularly called warm, because the heat of the body is maintained, or prevented from being dissipated, by the non-conducting quality of the materials used for clothing.

Cool dresses imply that the textures used are good conductors of heat, and therefore tend to lead away the heat from the body. Warm dresses, again, imply that the textures are bad conductors, and therefore tend to preserve the heat of the body.

11. *Explain the formation of dew and hoar-frost.*

Dew.—After sunset, the earth radiates the heat she has received during the day, and as the night advances, the cooling down becomes such as to affect the superincumbent air. That air becomes also chilled, and is no longer able to hold its aqueous vapour in suspension; condensation ensues, and the watery particles are deposited on the ground. The deposition is greatest on grass and foliage, because of their high radiating powers. A cloudy night or a windy night is unfavourable to its production.

Hoar-frost.—If it so happen that, after dew is deposited, the temperature of the air sinks below the freezing-point, then the watery particles are frozen, and appear as hoar-frost.

12. *Describe how the common mercurial thermometer is made and graduated.*

Construction.—A glass tube of small and uniform bore, with a bulb at the end, is taken, the other end being opened out into a cup-like shape. The bulb is then held over a flame, and when the air is sufficiently expelled, pure mercury is poured in. The bulb is again heated so as to expel the air, and another quantity of mercury is poured in, and so on until a sufficient quantity is introduced. When this is done, the bulb is once more heated until the mercury flows over at the top of the tube; then the little cup is removed and the tube hermetically sealed.

Graduation.—To graduate the instrument, it is plunged successively into ice-cold water, and into steam escaping from boiling water; the levels of the mercury at these temperatures are then marked 32° and 212° respectively. The space between these points is afterwards divided into 180 equal parts, and numbered accordingly.

13. *You place one hand in mercury, and the other in water, both liquids having the same temperature ; the hand in the mercury feels colder than the hand in the water—why ?*

The human body is in general at a higher temperature than surrounding objects, and as the tendency of heat is to pass from the hotter body to the colder, that body feels the coldest which conducts heat the most readily. Now mercury *conducts* heat much better than water; there is therefore more heat abstracted from the hand by the mercury than by the water in a given time ; hence the difference of sensation. Were both fluids hotter than the hand, the reverse sensation would be the case.

14. *Describe the heating powers of the various colours of the solar spectrum.*

The solar spectrum is made up of seven different colours, viz., red, orange, yellow, green, blue, indigo, and violet. Commencing from the violet end, there is very little heat perceptible ; the heating power, however, gradually increases towards the red, a little *beyond* which is found the maximum heating power. The heating powers of the different colours are usually shown by a curve above the spectrum, which gradually rises, until it attains its greatest height towards the red, a little beyond the visible spectrum, thereafter gradually falling, till it eventually disappears.

WILLIAM COLLINS, AND CO., PRINTERS, GLASGOW.

PUTNAM'S HANDY-BOOK SERIES.

1. THE BEST READING. A Classified Bibliography for easy reference. With Hints on the Selection of Books ; on the Formation of Libraries, Public and Private ; on Courses of Reading, &c., &c. 12mo, cloth, $1.25.
"We know of no manual that can take its place as a guide to the selection of a library."—*N. Y. Independent.*

2. WHAT TO EAT. A Manual for the Housekeeper: giving a Bill of Fare for every day in the year. Cloth, 75 cts.
"Compact, suggestive, and full of good ideas."—*Many Housekeepers.*

3. 'TILL THE DOCTOR COMES; AND HOW TO HELP HIM. By George H. Hope, M.D. Revised, with Additions, by a New York Physician. Cloth, 60 cts.
"A most admirable treatise; short, concise and practical."—*Harper's Monthly. (Editorial.)*

4. STIMULANTS AND NARCOTICS ; Medically, Philosophically, and Morally Considered. By George M. Beard, M.D. Cloth, 75 cts.
"One of the fullest, fairest and best works ever written on the subject."—*Hearth and Home.*

5. EATING AND DRINKING. A Popular Manual of Food and Diet in Health and Disease. By George M. Beard, M.D. Cloth, 75 cts.
"We can thoroughly commend this little book to every one."—*N. Y. Evening Mail.*

6. THE STUDENT'S OWN SPEAKER. By Paul Reeves. Cloth, 90 cts.
"We have never before seen a collection so admirably adapted for its purpose.—*Cincinnati Chronicle.*

7. HOW TO EDUCATE YOURSELF. By Geo. Cary Eggleston (Editor *Hearth and Home*). 75 cts.
"We write with unqualified enthusiasm about this book, which is unfailably good and for good."—*N. Y. Evening Mail.*

8. SOCIAL ECONOMY. By Prof. E. Thorold Rogers (Tooke Professor of Economic Science, Oxford, England), Editor of "Smith's Wealth of Nations." Revised and edited for American Readers. Cloth, 75 cts.
"We cannot too highly recommend this work for teachers, students and the general public."—*American Athenæum.*

9. HINTS ON DRESS. By an American Woman. Cloth, 75 cts.
"This little volume contains as much good sense as could well be crowded into its pages "—*N. Y. Mail.*

10. THE HOME. WHERE IT SHOULD BE, AND WHAT TO PUT IN IT. Containing hints for the selection of a Home, its Furniture and internal arrangements, &c., &c. By Frank R. Stockton (of *Scribner's Monthly*). Cloth, 75 cts.
"Young housekeepers will be especially benefited, and all housekeepers may learn much from this book."—*Albany Journal.*

11. THE MOTHER'S WORK WITH SICK CHILDREN. By Prof. J. B. Fonssagrives, M.D. Translated and edited by F. P. Foster, M.D. A volume full of the most practical advice and suggestions for Mothers and Nurses. Cloth, $1 25.
"A volume which should be in the hands of every mother in the land."—*Binghamton Herald.*

12. HEALTH AND ITS CONDITIONS. By James Hinton, author of " Life in Nature," &c. 12mo, cloth, $1.25.
"A valuable treatise on a very important subject."—*Louisville Recorder.*

13. WHAT IS FREE TRADE? By Emile Walter. 12mo, cloth, 75 cts.
"The most telling statement of the principles of Free Trade ever published."—*N. Y. Nation.*

9 783743 392595